SS GREAT
BRITAIN

SS GREAT BRITAIN

BRUNEL'S SHIP, HER VOYAGES, PASSENGERS AND CREW

HELEN DOE

AMBERLEY

Dedicated to the staff and volunteers of the SS Great Britain Trust, past and present

Jacket, front: The *Great Britain* in Table Bay, 1852, by John A Wilson. (SS Great Britain Trust)

Page 2: The *Great Britain* with her six masts by Keith A. Griffin. (SS Great Britain Trust)

First published 2019

Amberley Publishing
The Hill, Stroud
Gloucestershire, GL5 4EP

www.amberley-books.com

Copyright © Helen Doe, 2019

The right of Helen Doe to be identified as the Author of this work has been asserted in accordance with the Copyrights, Designs and Patents Act 1988.

ISBN 978 1 4456 8451 2 (hardback)
ISBN 978 1 4456 8452 9 (ebook)

British Library Cataloguing in Publication Data. A catalogue record for this book is available from the British Library.

Typesetting by Aura Technology and Software Services, India. Printed in the UK.

CONTENTS

FOREWORD

In September 1838, it was announced that the Great Western Steam Ship Company was to build a sister ship to its first transatlantic paddle steamer. The directors ordered a new *Great Western*, but what they got was entirely different. The new ship when launched in Bristol five years later was the world's first iron-built screw-propelled passenger liner, and of giant proportions. Initially known as the *Mammoth*, this was Brunel's SS *Great Britain*, and of all the far-reaching engineering adventures for which I. K. Brunel is rightly famous, this ship is probably his masterpiece.

This may seem a bold statement, but it certainly seems to stand up to the facts. Not only would ships and their technology never be the same again, but the volume and distance of their impact on burgeoning world trade routes would lead to a process of globalisation and mass migration which continues ever more strongly today. The human world was shrinking, and for the first time people and cargoes could travel to and fro on their business speedily, safely and reliably.

Wonderfully, this ship has survived the rigours of some 32 voyages around the globe and a million miles at sea during her long working life. She lies today preserved in the very dry dock built to construct her in the heart of Bristol as a living museum and heritage site dedicated to the public understanding of her history and significance, and to stimulating the enthusiasm of young people for the science and engineering of the world around us.

Dr Basil Greenhill, one-time Director of the National Maritime Museum in Greenwich, London, wrote: 'It is impossible to imagine a more important survivor of our "heritage", in terms of her significance to the industrial, economic, and social development of Britain.'

It has been said that she is somewhat less known on the international stage than her later cousin – Brunel's giant *Great Eastern* of 1858. There are perhaps two reasons for that.

Firstly, many of the twentieth-century maritime history books failed to see beyond the dramatic 11-month grounding of the SS *Great Britain* on a beach in Northern Ireland in 1846 on only her fifth voyage towards New York. Indeed the story is then often abruptly stopped there; failing to notice not only (and very significantly for shipbuilding technology) that this huge iron ship did not break up in the winter storms, as most previous ships would have done, but also that she subsequently went on to sail regularly and quickly to and from Australia with thousands of passengers and migrants each way, dominating that passage for many years with her speed, comfort and reliability.

And secondly, it is worth noting that the *Great Britain* was born at the time of the very infancy of photography – Charles Fox-Talbot took one of the earliest ever negative photographs of a ship when capturing her lying alongside while fitting out in Bristol Docks in 1843/4. By contrast the *Great Eastern* has enjoyed a heavily and well-illustrated life which naturally lends itself to the best illustrated history books and documentaries, perhaps giving her an international profile that her predecessor deserves a little more.

Ewan Corlett started the task of rescuing both the ship and the reputation of the SS *Great Britain* in the late 1960s, and documented her well in his seminal, if somewhat technical work, *The Iron Ship*. It is now nigh on 50 years since the SS *Great Britain* was towed back to Bristol and preserved for an international public to enjoy, and Dr Helen Doe has therefore done us all a great service in bringing together the whole story of the magnificent ship.

Now the restored ship is the centrepiece and highlight of Bristol Docks, and the leading visitor attraction and museum in the city, as voted by the public. She is a piece of tangible history and a physical link with the past available for everyone today.

The personal histories of her many passengers and crew, the breakthrough combination of cutting-edge technology envisioned by I. K. Brunel, and the international impact and legacy of world travel that the ship symbolises are brought together in a modern and accessible treatment. This is the first book on the history and significance of Brunel's masterpiece in the 21st century, and is greatly to be welcomed and enjoyed.

Dr Matthew Tanner MBE
Chief Executive, SS Great Britain Trust
Great Western Dockyard, Bristol 2019

INTRODUCTION AND ACKNOWLEDGEMENTS

The *Great Britain* is an iconic ship and, together with HMS *Victory* and the *Cutty Sark*, ranks highly in our national consciousness. I am privileged to be a trustee of the SS Great Britain Trust and am therefore a regular visitor to the ship, a sight that never fails to delight me. The thousands of people who visit her each year testify to her popularity, over 200,000 in 2018. Ewan Corlett, who did so much to ensure that the ship returned from the Falklands, wrote *The Iron Ship* in 1975, and it was subsequently republished by the SS Great Britain Trust in 2012. Written by a very knowledgeable naval architect who knew and loved every part of the ship, it is the classic technical book. My intention with this book is to put the ship into a wider context, and in writing today I have access to sources that were not available to Ewan Corlett.

Over the years the Trust has collected a considerable amount of material and in the Brunel Institute it now holds the world's collection of Brunel archives, much of it on loan from the University of Bristol. In addition, many volunteers over the years have continued to explore and to research her career, and a lot of their findings are held in the David Macgregor Library within the Brunel Institute. Jean Young's hard work was key to this collection and her skills as a librarian were put to good use, sourcing material on the ship from archives and libraries far and wide. The most recent example of this continuing work is the online Global Stories project, freely available via www.ssgreatbritain.org. It took many hours of volunteer work to transcribe all the passenger lists and all the existing crew agreements onto a database. Through the kind offices of Nick Booth, curator of the ship, I have been able to access this database, which has in excess of 36,000 names and associated data. Additionally, the Global Stories project utilised other information from diaries and information from relatives of passengers and crew on board to provide vignettes of some of the people who travelled and worked on the ship. It is a fascinating project and continues to grow.

Another area of my research has been to use the many digital versions of 19th-century newspapers now available online. These are an incredible treasure trove of information. They do have to be treated with care since there is no presumption of accuracy in the reports, but they do allow the researcher to find information that simply would not be available elsewhere. They provide a window into the information available for the Victorians and their attitudes and interests. Through these newspapers new passenger and crew information has emerged, together with more information on the companies. Company information is often hard to find, particularly if they were private companies.

Through these various sources and many others, this book looks at the people on board, both crew and passengers, but also looks at the companies who owned the ship. Their decisions, their financial success or failure, all had an impact on the *Great Britain*. In addition, there were various other factors such as the political scenario in which these companies were operating. War and mass emigration determined her later career. When writing about Brunel's first ship, the *Great Western*, I learnt a great deal about the Atlantic trade and the growing city of New York. In this book I have had the pleasure of learning much more about Australia and the many people who went out on the *Great Britain* as emigrants to start a new life. There are still many more avenues to research, so many more things to more to discover.

In writing this book I have been indebted to the many volunteers at the SS Great Britain Trust, some of whom I have never met but who have left their work in the library. Of the current volunteers I must thank Malcolm Lewis and Peter Revell for giving me access to their research notes and Alistair Roach for checking dates. Mike Hinton similarly allowed me to use his notes on his Crimean research. Nick Booth and Mollie Bowen have been incredibly helpful in providing access to documents, books and transcripts. I was delighted when Matthew Tanner, the Chief Executive of the Trust, agreed to write the foreword to this book. He was simply the only person who could do this. His involvement with the *Great Britain* goes back to 1997 when he was appointed as the first professional curator. Since then, through his energetic leadership, his passion for the ship and Brunel and his creative vision of the future, the SS Great Britain Trust has expanded and had remarkable success in a difficult sector. The Trust has rightly won over 50 awards during his tenure and I am most grateful that he was prepared to write these words.

Dr Graeme Milne of the University of Liverpool provided me with some of his research, Dr James Muirhead allowed me to use a copy of his thesis and Dr James Boyd, Research Fellow at SS Great Britain Trust, kindly read a final version. My late friend Mike Stammers appears in this book on a regular basis. His work on the Black Ball line to Australia was one of the last books he published and he was kind enough to send me a copy. His knowledge was extensive and he is much missed. More locally to me, Kate Taylor was of immense help in beavering away to research some of the stewardesses and some of the stories linked to them; I really am very grateful to her for that.

My regular support team as ever has been around. Thank you to Kathleen, Gillian, Ruth and Nikki, who keep my life on track. My small supervisor, Jenna, has been unceasing in her efforts.

A particular thank you goes to Joanna Thomas, maritime curator at the SS *Great Britain*. She is currently also a PhD student studying at Exeter and a new mother, so a very busy lady. Joanna very kindly read my early drafts and we have had some most enjoyable discussions and she has added considerably to my knowledge. Any errors in the book are wholly mine and do not reflect on those who so kindly helped me.

My big supporter is my husband Michael, who has coped with the usual fraught time as the book comes to fruition. None of my writing would be possible without him.

I

'YOUR OTHER SHIP': A SECOND VESSEL

The Great Western Steam Ship Company of Bristol had successfully established the first steamship line across the Atlantic with just one ship, the *Great Western*. At the time of her launch in 1838, this wooden paddle steamer was the largest ship to be built. The ship was a speculative effort, driven by new engines and setting out to disprove the vocal critics who did not believe it was possible for a ship to steam continuously across the Atlantic. The *Great Western* had proved the critics wrong and established a regular two-week crossing, even against strong headwinds and, with no direct competition in its early years, was financially very successful. But any company establishing a regular ocean route could not rely on just one vessel. A second was needed, to provide more sailings and to ensure backup and continuity. It had been the company's intention from the beginning, on the advice of their consulting engineer, Isambard Kingdom Brunel, to have at least two vessels on the crossing.

The company had the timber; they had the expertise in their shipbuilder, William Patterson, at whose yard the first ship had been built; and they good relations with Maudslay, Sons & Field, who had designed the engines. A direct copy of the *Great Western* could be speedily and effectively built and launched, but it was not to be. The ship that would eventually be launched, the *Great Britain*, was to be different in almost every aspect and would take six years to be realised. It was 1845 before the Great Western Steam Ship Company had two ships across the Atlantic. So what happened?

Even as the proprietors, as the shareholders were known, celebrated their great triumph in crossing the Atlantic, marine steam was still in its experimental stages as a technology. Sail would dominate the shipping registers and fill the ports until the 1880s, and sailing ships continued to be built into the twentieth century, even if most of the technical refinements in sail peaked in the 1850s and 1860s. Huge steel sailing ships would still continue to be improved for the South American nitrate trade even in the early twentieth century.[1] The arrival of the steamship had a massive impact on the world of shipping in the 19th century and this was felt not just at sea but also on land, changing occupations, providing new business opportunities, speeding communications and information exchange, changing the architecture of many ports and enabling Britain to expand its influence across the globe. But while it first came to national attention in the early years, it took many more decades for steam to replace sail. Steam was enthusiastically adopted on rivers, canals and estuaries and short sea crossings. There, coal supplies and specialist engineering skills could be speedily supplied when needed. Crossing oceans, particularly on very long-distance routes, proved more challenging to any regular operation.

To talk of steam alone oversimplifies the technological innovations required to make a steamship line successful and profitable. The Great Western Steam Ship Company's ships, the *Great Western* and what would become the *Great Britain*, provide illustrations of the many interlocking factors involved in this process and were among the first attempts to manage a complex process of innovation. Technological change does not happen in isolation. Brunel and his fellow engineers were only too aware of this, and collaborated to a high degree. Steam engine design, boiler efficiency, skills in iron hull construction, the design and manufacture of propeller shafts and housing, hull and rigging design were all being tested and developed in different places and at different times. 'Steam shipping as a whole could only progress as fast as the slowest element in the array of new technologies, which had to work in concert.'[2] They all came together in one ship which would lead the way and they all required the determination of the people involved, from engineers to company directors and the long-suffering investors. To make the experiments succeed, the business required crews, brokers, agents, cargoes and paying passengers.

Much has been written about the *Great Britain*'s technical innovations as the first iron-hulled, screw-propelled steamship, leading as it did to many future generations of steamship. It is not the intention of this book to revisit all the technical arguments; Ewan Corlett and others have done that in detail.[3] Here, the intention is to look at the wider aspects of the ship, the people who managed her and were carried on board, and the companies that owned her. She was, of course, not the only iron ship, nor the sole ship to carry passengers to Australia; nor was Brunel alone in his endeavours.

It is also important to consider *where* technological change was happening, as the political, social, cultural and economic environment is relevant to the process, just as the various individuals and firms involved in the ship's creation were part of a broader commercial web.[4]

Once the *Great Western* had safely crossed the Atlantic, Brunel and the building committee turned their attention to the sister ship. It had always been the intention to have two vessels on the line to provide a regular service and ensure backup. The initial plan had been a simple one: build a ship identical to the *Great Western*. If they had remained with this plan and spent their money on that sister ship, the future of the Great Western Steam Ship Company might well have been very different. Their reliability would have improved, they would have had a steady income and might well have had a better chance of securing a government subsidy from carrying government mail and despatches. But the earnings of the *Great Western* as she ploughed steadfastly across the Atlantic, and the monies raised through the shares of the company, went on experimental grand designs, as Brunel and his colleagues saw themselves as pioneers in Atlantic steamship technology.

The company's building committee consisted of the Bristol shipbuilder William Patterson; the company's managing director,

Christopher Claxton; another director, Thomas Guppy; and their consulting engineer, Isambard Kingdom Brunel. These four had worked successfully on the predecessor, the *Great Western*, which had been built in Patterson's yard in Bristol. An early decision to make the second ship larger caused difficulties. The larger size, it was thought, would bring greater profit, with more passengers and more cargo space. The design took some time, and according to Brunel's son 'each succeeding drawing increased the size proposed; at length the fifth design, showed a ship of 3,443 tons burden', far, far larger than the *Great Western*, which was 1,340 tons when launched.[5]

Emboldened by the success of their first ship, and aware that Patterson's yard could not build the new, larger vessel, the company took the decision to invest in their own shipyard. Additionally, they needed to consider the requirements not just for building a new ship but for maintaining their existing vessel. They explained the decision to the company's proprietors:

> To remedy in future the great inconvenience, expense, and labour, which were incurred in building a yard of limited space, and also to hold your stock of timber (which is equal to the construction of a steamer of more than 200 tons) together with your ways, planks, scaffolding, stages, and standards; and, for the more permanent operations of the company, they have taken, on a lease of 21 years, determinable by the company at seven and 14, most convenient premises on the lower part of the Bristol floating harbour.[6]

This was a significant statement of their intention to be a serious steamship line, and the step was taken before many other competitors had reached this stage. But being first out of the blocks is not always a guarantee of winning the race.

Timber was the accepted material for all shipbuilding, and a large amount of timber was bought by the company to build the next paddle steamer. Indeed, they had sufficient timber to build not just one further vessel, but two. But despite this purchase, within months there was a change. In October 1838, an early iron-built coastal steamer, *Rainbow*, visited Bristol. Iron had been used for the occasional small vessel since 1819, and it was still being evaluated as a material. The building committee of the Great Western Steam Ship Company examined the *Rainbow* and pondered the practicality and benefits of iron ship construction. One of the benefits was that an iron ship could carry more cargo than a wooden ship of the same dimensions. The thinner shell plating and the strength that came from the iron beams and frames could increase the internal space by up to 20 per cent. Another considerable benefit was the ability to build ships of a greater length. The *Great Western* was the largest ship of its time when it was launched, but it had to be very thoroughly strengthened and structured in such a way as to mitigate the inherent weakness of a long ship made of timber.

Brunel noted other benefits of iron construction, such as its comparative safety in ice.[7] He listed qualities including the lack of dry rot and he rather optimistically believed, based on what can only have been limited evidence, that there would be no vermin.[8] Deciding to build in iron, and having decided to build the ship themselves in their newly acquired dock, was an ambitious and bold move. There were few shipbuilding bases constructing iron vessels. Highly skilled shipwrights were well accustomed to wood, but iron needed to be cut and drilled by hand to a very high degree of precision, and the riveting of the plates also required new skills. These would need to be employed. The scale of the ship was to be considerably greater than any other iron ship yet built. The iron ship *Royal Sovereign* was built in 1838, but the proposed ship would be two and a half times wider, twice as long and eight times as heavy. The company was working at the edge of known technology; there was no published literature on the construction of iron ships.[9] At the company's annual general meeting in March 1839, the directors announced their change of building material and justified their approach to the second ship as being 'unusual caution in the construction and an increased conviction that the nearest possible approach to perfection must be kept in view'.[10] It was a very Brunelian statement.

The engines were the next big procurement matter. As with their first ship, three firms were invited to tender for the engines: Maudslay, Son & Field, who had built the engines for the *Great*

Western; Hall; and Seaward. Eventually, after some correspondence, the decision was between Maudslay and one Mr Humphreys, whose patents for trunk engines were manufactured by Hall. Brunel was very clearly in favour of Maudslay, and on more than one occasion persuaded the Great Western Steam Ship Company directors to delay their final decision while waiting for more details from Maudslay, as in the following letter from Brunel to Claxton on 7 May 1839:[11]

> My dear Claxton
> What a state you are in if the Great Western is not arrived and how busy you are if she is – yet what I have to say is worth your hearing.
> Maudslay's have not yet secured their patent, but I have seen their model. I must think well over it before I express any decided opinion upon its merits but this I can safely say, it's decidedly new and upon the face of it there is enough to make it worth our while to wait a week or 10 days by which time they will be prepared with estimates et cetera–
> Yours sincerely I K Brunel[12]

Brunel's association with Maudslay, Son & Field was a long one, dating from his father's connection with the firm in the very early days of marine engines. The firm, based on the Thames, was one of the best in the world, but their eventual tender was considerably more expensive than that of Humphreys. In Brunel's view, Humphreys had considerably underestimated the cost of building the engines. In addition, Hall's directors were somewhat reluctant to get involved since supplying engines to Humphreys' patent for such a large ship involved the purchase of specialist, very large and presumably expensive tools. They urged the Great Western Steam Ship Company to consider manufacturing their own engines.[13]

This was a big step, and would change the nature of the company from shipowners to complete shipbuilders and marine engineers. Some shipping companies preferred to remain as shipowners, placing their ships in a shipbuilder's yard for repair and maintenance. Traditionally, the roles of shipowner and shipbuilder were separate, as had been the case in the construction of the *Great Western*, and Patterson and Brunel had worked in partnership with the engine builders. Steamships brought together the separate skills of the engineer and the shipbuilder, which was not always a harmonious match. To combine hull, engines, boilers and paddles with a satisfactory result was no easy task. In 1861, the shipbuilder Murray, with many years of experience behind him, described this combination as 'one of the most difficult problems of modern engineering demanding at once the theoretical attainments of the natural philosopher and the laboriously acquired knowledge and shrewd sagacity of the practical mechanician'.[14] Naval architecture was not yet a recognised profession. The Institution of Naval Architects was not formed until 1860 and was largely ignored by mercantile shipbuilders until the late 19th century. Trial and error seem to have predominated, with many ships having to be corrected later.[15]

Joshua Field, of Maudslay, Son & Field, when speaking to a Parliamentary Select Committee in 1847, was asked his view on John Scott Russell, a well-known name in steamship circles who had carefully examined the action of a ship through water and invented the waveform design for ships. Field was asked if Scott Russell was a shipbuilder or a practical engineer. Field replied in a somewhat dismissive way that he was a 'literary man'.[16]

The idea of the company manufacturing its own engines was defended to the proprietors at the next annual general meeting. The directors explained they had looked at the capabilities of manufacturing businesses in Bristol and came to the conclusion that the company would be its own manufacturer, 'as many of the most successful Steam Navigation Companies already are'.[17] There were several shipbuilding companies based in Bristol in addition to Patterson. Among them were George Hilhouse & Co., who were still building sailing vessels. Lunel and company were working on steamships but their vessels were considerably smaller, averaging 500 tons.[18]

After a meeting with Peter Maze, the company chairman, Brunel, in his letter dated 12 June 1839, compared the plans and estimates from Maudslay and Humphreys and, while he tactfully declared himself happy with both sets of plans, the main difference in his mind, he wrote, was whether the engines were built elsewhere or in-house. He mentioned the 'advantages to be derived from the responsibility and experience in all the details of a first rate manufacture, and to which I attach very great value, particularly in the early proceedings of a company like ours'.[19] He went on to say, 'My only fear would be that of the risk of the undertaking being too great for a newly formed establishment.' He recognised that once the expensive tools and shops and the establishment had been completed, 'it would be rendered fully competent in point of means to continue the manufacture of engines for others, and to keep up with repairs of any number of engines which the company are likely to have at work.' He also pointed out that rather than being against the company doing any manufacturing, he simply felt it was the wrong stage at which to do it and that when it was a more mature organisation 'we might no longer depend entirely upon the engineering talents of one director... Or upon the health of one superintendent who as yet is alone in possession of all our plans and ideas and at present is alone capable of carrying them out.' He finished by saying that the cases were evenly balanced but that he was in favour of using the services of 'an old established manufactory, and the great relief of responsibility and risk obtained by contracting for the whole work'.[20]

The directors, despite the strong advice to the contrary from their consulting engineer, Brunel, decided to hire Francis Humphreys and have his trunk engine design built by the company in their own workshop.[21] Brunel, as the company's consulting engineer, was never happy with Humphreys and their relationship was fractious. Building began and the keel was laid in the company's building dock in December 1839. There had been problems in constructing the dock, and this delayed matters, but all seemed set for completion in 1841.[22]

The dimensions for the second ship were now set, the construction material had been decided and the engines were being built, all on a single site leased by the company in Bristol. The anxious proprietors needed to be appeased, and at the next annual general meeting the directors were at pains to show the full advantages of their own workshops. The winter maintenance of the *Great Western* had been carried out in the company's dock and both Patterson, as chief shipwright, and Humphreys, as engineer, came in for high praise. Had it not been for 'your own yard and workshops, and the floating shop, which the great distance or ship's berth in the river rendered it imperative to fit up, sailing on her appointed day would have been out of the question'.[23]

In referring to 'your second vessel' there was much flattery of proprietors and earnest wishes to ensure they understood fully and supported the significance of the work being undertaken,

> ... aided by the same scientific and zealous cooperation, which enabled them succeed so signally in the *Great Western*, they have given considerable progress in her construction. Both reason and experience taught them, that the continuance of the preference hitherto given by passengers to your company, would alone be insured by maintaining your already established superiority over other parties engaged in similar undertakings.
>
> It was evident also that great capacity and power, by which comparative economies secured with the highest efficiency, were to be the means through which your position was to be held, and the acquisition of the best constructed iron vessel, and of the most powerful marine engines yet made, has required and received the gravest consideration. The happy results which had attended your construction of the *Great Western*, the necessity of improving on the ordinary mode of building, and the command of the same energy and scientific knowledge, which had before been exercised on your behalf, left no doubt on the minds of your directors, that your own yard would best afford the means of both economy and excellence.[24]

There was reference to the length of time it had taken to decide on the proportions of the vessel, and the directors demonstrated to the proprietors the final model with the cheering statement that 'a large proportion of her frame is up, a considerable part of her bottom is riveted in place, and several of the most difficult pieces of forged work are completed.'[25]

It was confirmed that to reduce risk and insurance cost the ship would not be launched in the traditional way but floated out of a newly excavated dock after the installation of the boilers and machinery. Having such a dock 'for the purposes of docking and repairing your vessels will, with the additions which will be gradually made, render your company perfectly independent, whichever increase may take place in steam navigation from the port'.[26] This would imply the company was soon to be future-proofed – a rather rash statement considering the vast changes they had already overseen in such a short time.

The iron for the construction, they confirmed, was purchased from the Colebrookdale company on satisfactory terms. The subject of engines for the iron ship had involved many estimates from 'the most eminent manufacturers' and reference was also made to the 'mass of data', as they called it, from the observations conducted during the voyages of the *Great Western*.

From the very first crossing, Brunel had been collecting data about the performance of the ship. He regularly put his assistants on board the *Great Western* to monitor performance and take detailed notes of how the engine performed and how the ship itself performed at sea. He took a very scientific approach to collecting data. The directors were also able to report that they were already well advanced in construction of the trunk engine.[27]

After the good news came the difficult part: further finance was needed to progress the ship and the machinery. The directors proposed that temporary loans were taken out on the 'ships, buildings etc.'. This proposal was not taken up by the proprietors. The directors had extended the lease of the premises on the harbour to 31 years, and all buildings and fixtures remained the property of the company at the end of the lease, with the option being reserved by the landlord to take them at a valuation. The fitting-out of the buildings included large lathes, planing machines and other items of machinery. Particular note was made of the assistance of Mr Brunel and his getting the opinions of the 'first mechanics and machine makers'. His 'assistance at every spare moment in it and at any personal inconvenience himself has been at the service of your directors'.[28]

With all of this investment in buildings, tools, machinery and skills, the directors hinted that the proprietors might also look at manufacture and repair for other parties. They pointed out that 'in this new and most important field … no existing establishment in Great Britain has capabilities for making engines of the dimensions of those of your new ship, and there is no other in which a first-class steamship, with engines, can be built, and for which she can be turned out, in a finished state, with equal economy.'[29] This again was a new thought, broadening their remit, but as a company they were bound by a deed of settlement.

The deed meant that they could only use the company's funds for the 'purposes specified', which presumably did not include building and repairing ships for other parties. The proprietors, they suggested, should agree to broaden the deed. There was also at this stage a suggestion that the company had the opportunity to apply to the Privy Council for a charter, limiting 'your responsibilities and affording other benefits to your company'. The company had been established as a joint-stock company but was not incorporated, as this required Parliamentary approval and was a very expensive process. Instead, they took the unincorporated route and established the company by deed of settlement. The process they had undertaken involved establishing joint-stock, but the stock was held by trustees who committed to observe the company's deed of settlement.[30]

At this stage also it is worth noting a debate on insurance, something that would later come back to haunt them. Not all ships were insured in London as some owners found Lloyd's of London too expensive and formed mutual insurance companies based in local ports. Other owners simply did not take out any insurance. New and experimental ships were of concern to Lloyd's, and Lloyd's Register had difficulty classifying them. The directors of the Great Western Steam Ship Company had taken out insurance on their existing ship to the 'extent of £50,000. Which for the current year has been reduced to the rate of six guineas percent, with a return of 10 shillings percent for each unemployed month.' This they considered an 'excessive charge of your ship, they'll be glad to know the views of the proprietary on this subject'. Mr Fripp, seconded by Mr Moxham, proposed 'that at the conclusion of the present policy of insurance, the directors take into their serious consideration the propriety of its not being renewed, but that it should be left to the option of individual proprietors to insure if they thought proper'.[31]

In May 1840, another ship visiting Bristol was to have a dramatic impact on the embryonic *Great Britain*. The screw steamer *Archimedes* promised a wholly different method of propulsion. The idea of using a screw propeller had been demonstrated on the Hudson River in America, and in the 1820s Marc Brunel, Brunel's father, who was a noted engineer himself, conducted model experiments in a tank. It was Francis Pettit Smith, however, in May 1836 who took out patents for the screw propeller, and Captain Ericsson from Sweden took out another for his propeller in July that year.

A big difference between these two patents was that Smith, by background a farmer from Hendon, placed his screw propeller in front of the rudder, which was a more effective position.[32] John Ericsson's screw propeller was much more complicated in its design. Both inventors built models and tested them, with Smith utilising his farm pond. Smith progressed to a 34-foot experimental vessel on the canal at Paddington. While paddlewheels could provide stability in many circumstances, there were limitations on the high seas. One factor was that as the coal stocks diminished the ship rose higher out of the water and the paddle wheels had greater difficulty in driving the ship forward; similarly, heavy seas might cause one wheel to come out of the water. The propeller did not have that problem. The Ship Propeller Company was formed in 1839 to exploit Smith's patented location for the screw propeller. It was well funded and included John and George Rennie, who were engineering contractors to the Admiralty. They built the 200-ton *Archimedes* to prove the propeller concept. The ship was sent on a trip around the coast of Britain after her initial trials to publicise the screw and arrived in Bristol in May 1840.[33]

Guppy, Claxton, Patterson and Brunel all visited the *Archimedes* and Guppy travelled with the ship to Liverpool to make further observations. Such was the impact and possibility of this new form of propulsion that the directors were persuaded to abruptly halt the construction of the *Great Britain*'s engines and those parts of the hull that might be affected by such a major change. This cannot have pleased Humphreys, who was now temporarily redundant. The Great Western Steam Ship Company then hired the *Archimedes* for several months from the Screw Propeller Company in order to carry out their own tests with different types of screw propeller.[34]

With the agreement of the owners of the *Archimedes*, Brunel, assisted by Guppy and Claxton, carried out various experiments.[35] Brunel was impressed by the merits of the propeller, but he thought more work needed to be done on both the size, shape and the fitting of the screw.[36] Brunel was particularly interested in how to transmit the power from the engine to the propeller. Brunel wrote to Claxton:

A short time back Barnes, who was with us on the *Archimedes* told Phipps that he had considered the results of our experiments had made the screw better than the common paddlewheel... But that taking into consideration all the advantages of the screw, it was better than any paddle and that he had no doubt it would soon supersede the paddle. I never heard this till today. Phipps is positive of the whole. This is satisfactory. But except to Guppy or Bright [Great Western Steam Ship Company directors] do not mention it.[37]

As a result of all the testing, Brunel's lengthy – and now famous – report to the directors of the Great Western Steam Ship Company on the advantages of the screw propeller over paddle wheels was presented in October 1840. It was detailed and complex, as the board of directors had to be convinced that this tiny screw propeller would be as effective, if not more so, than the vast and impressive paddlewheels. Brunel explained that although 'there is little or no appearance of any rotary motion in the sense of the water it is not put into rapid motion as in the case of the paddle... Comparing this with the violent displacement of the water by the action of paddle boards even under the most favourable circumstances I no longer feel surprised.'[38]

Captain Claxton's son, Berkeley, was an engineer and draughtsman working for Brunel and he went on many voyages on the *Great Western* to measure the pitch and the roll in the rise and fall of the sea at the stern. This enabled Brunel to point out that the operation of the screw was unaffected by the trim or rolling of the vessel and allowed free use of sails.[39] It saved a considerable amount of weight on the ship and a made for a better and simpler form of vessel. Demonstrating his considerable theoretical knowledge of hydrodynamics, the results of the many experiments demonstrated by the *Archimedes* and the data from the *Great Western*'s performance on the Atlantic, Brunel's report firmly recommended that the new steamer should be adapted for propeller. He stated at the beginning, 'My opinion is strong and decidedly in favour of the advantage of employing the screw in the new ship... I am fully aware of the responsibilities I take upon myself by giving this advice.' He personally attended the board meeting in December 1840 to explain his report and, after some debate, the directors resolved to adopt screw propulsion.

So, in December 1840, came the second major change to the new ship, as the directors agreed to adopt screw propulsion and Francis Humphreys resigned from the company. Reluctant to introduce someone totally new who would have taken a long time to become acquainted with the machinery of the ship and the new propeller, the directors sought the views of Brunel as consulting engineer. His advice was to obtain the services of Guppy, a director of the company, who would become supervisor and manage the manufactory on a full-time basis. Guppy had been closely involved for some time, having been a key figure on the Building Committee. Mr Harman, who had previously worked for Maudslay, was appointed as engineer in chief on Guppy's recommendation. The trunk engine design was no longer required so Brunel turned to his father's early design for a marine engine. The marine engine was the largest marine propulsion unit of its day.[40]

Altering the *Great Britain* from paddle steamer to screw propulsion involved several design alterations to the whole, but this did not seem to cause any problems.[41]

By March 1841 the directors had to explain to the shareholders yet again a major change to the new ship, which was now completely unrecognisable as a sister ship and was still far from ready. Good news was promoted strongly and the sagacity and benefits of their decisions underlined. 'The masonry of the various buildings and of the dock have stood the severe trial the past winter without showing a symptom of weakness. Your entire stock machinery is now in daily and active operation.' Having made the decision in the best interests of the company to 'get rid of the paddle in your iron ship and to substitute a propeller upon the principle of the Archimedean screw', the work had progressed to effect the change and 'slight alterations required in her frame were easily affected and at a trifling cost.' Grateful thanks were again given to Mr Brunel as consulting engineer.[42]

Finance was still an issue and there was a reminder that the proprietors had themselves decided previously not to empower the board to raise temporary loans. Had this happened, the chairman said, 'the completion of your second ship in 1841 might have been effected, but this without unduly hurrying her work cannot now be accomplished.' It is noticeable that the directors always referred to their investors as proprietors rather than shareholders. But the directors confidently predicted that she would be ready for service by March 1842.[43]

Reference was made to the 'late Mr Humphreys', who had died shortly after resigning his post, and it was announced that the plan to expand the role of the company into a shipbuilding and repairing company for others was shelved. 'The work which they had at the time more immediately in view ... had passed into other hands and they were aware from a letter from several proprietors who objected to that change to the deed of settlement.'[44] Finally Mr Maze announced his retirement as chairman and his place was taken by Mr Robert Bright, the deputy chairman.[45]

Having placated the proprietors, the company continued its work on the ship, the progress of which was now becoming an object of considerable curiosity. to the delight of Captain Claxton: 'The sides of the *Great Britain* were scarcely visible over the walls of the yard in which she was building, when naval officers, ship builders, engineers, and philosophers from all countries began to seek admittance, and many have been the papers which in most languages have been written on the comparative merits of iron and wood as material shipbuilding.'[46] The company was only too happy to show off their ground-breaking vessel to the world.

2

'THIS EXTRAORDINARY VESSEL, THE WONDER OF THE WORLD'

The directors had promised that the ship would be ready by March 1842. Most of the shareholders were locally based, several were well-informed shipowners and most recalled that their first ship had its keel laid in June 1836, was launched one year later, and was completed and in service by April 1838. Their second ship, in contrast, which should have begun in early 1839, had its keel laid in December 1839 after the change to iron and now they had to announce it was still not ready. The March 1842 annual general meeting was deeply troubled and matters came to crisis point. The profits from the *Great Western* had not been as good as before, and part of that they attributed to the tragic loss of the *President*, a rival steamship, which they was believed had 'affected public confidence in transatlantic steam'.[1] Reference was made to the continuing efforts to get a grant from the government for carrying mail and dispatches and the heavy port charges in Bristol that caused them to consider moving the ship to Liverpool.[2] Unable to make further calls for finance from the shareholders, work was reduced during the winter, thus delaying the completion of the ship, now named the *Great Britain*, until spring of 1843.[3]

The proprietors were deeply concerned about their exposure to risk, and a committee of ten proprietors, chaired by George Jones, had written a report. The agitated proprietors were keen to divest themselves of liabilities and to raise capital to cover the escalating costs of the new ship. As an unincorporated company their risks were higher since their organisation was a mix of partnership and trust law. The dockyard was put up for sale in July 1842, unsuccessfully, and the *Great Western* was placed in an auction in October 1842. Neither action resulted in a sale.

Some shareholders took to the newspapers to express their frustration. 'I am one of the unfortunate shareholders and have now become so sick of the continued demands and want of money that I feel almost reckless of the consequences,' wrote one man and he pointed out that in August 1841 the cost had been just over £76,000, whereas now it had risen to £100,000. Some had wanted to reduce costs by reducing the weekly wages, but he felt that the opposite approach should be taken to incentivise the builders to finish it more quickly, thus ensuring the ship went into operation and began earning. Guppy, as the director in charge of the whole engineering works, had been under strong pressure and some felt his services should be dispensed with. The frustrated shareholder makes an interesting observation, which highlights the development of the profession of engineer at the time. 'If we are to finish the ship let us procure the additional assistance of a real practical industrious engineer in whose ability we may place confidence the short time we are to remain a company… I do not wish to be understood as hinting at any incapacity in Mr Guppy as engineer seeing that the title and diploma may be had even easier then by setting one's legs under the office table of a CE in chief for a couple of years… One cannot but think that the people of Bristol with the exception of two, are perfectly innocent of sleeping with one eye open: to me it appears they keep both shut.'[4]

This was a direct reference to Guppy's apparent lack of engineering background. He had never undertaken a specific apprenticeship, although he had tried in his youth to work with Maudslay. Instead his experience was more business-based.[5] This was not unusual in these early days – Brunel himself had an unconventional background, an apprenticeship to a clockmaker in France before learning engineering from his father. Corlett, when writing his book on the *Great Britain*, viewed this criticism from the shareholder from his perspective as a twentieth-century naval architect and commented that such criticisms were 'hardly fair. At a time of rapid change and shipbuilding it was not surprising that changes should be made and costs increased.'[6] That may be true from an engineering perspective, but the long-suffering investor had a good point. He had invested on the basis of promises of a good financial return, but all he was now hearing was delay and expense.

The dissatisfaction of the shareholders culminated in a heated extraordinary general meeting held in private on 11 November 1842, when it was finally agreed that the company should be permitted to borrow £20,000 to finish the ship.[7] In this meeting, there is no doubt that Robert Bright, as chairman, had been the key player in turning matters around and pacifying the meeting. He had saved the day for the *Great Britain*, but other company assets had to be sold and expenditure on the ship was reduced.[8]

A few months later, at the next formal meeting of the company in March 1843, the board reported a mixed financial situation. The shareholders' insistence on realising capital from the asset sale had so far been unsuccessful and the company had failed in its efforts to get a subsidy from the government in the form of a mail contract. Reference was made to the 'continued dis-arrangement of the monetary and commercial concerns of the United States, and general stagnation of business'.[9] On a more positive note, the move to operate the *Great Western* from Liverpool rather than Bristol had improved matters; costs were reduced and the ship was bringing home passengers who would not have joined in New York for Bristol.

The reduced weekly expenditure on the *Great Britain* had slowed progress.

She is however in a very forward state. The frame and hull are complete, the whole of the upper decks, as well as the decks of the forecastle, fore cabins and after cabins are laid and caulked; nearly the whole of the state rooms and other joiners work of these parts is finished; the fore hold, after hold and iron cold decks, before the boilers and abaft the engines, are nearly finished, the boilers and funnel are finished and fixed in their places, as are the cylinders, condensers, air pumps, and other weighty parts of the engines.[10]

The directors summed up the situation positively: 'Indeed, unless any unforeseen cause or delay should arise, she may be floated out within three months.'[11] Finally there was a heartfelt tribute to Mr Robert Bright, who had resigned his seat at the board due to a bereavement. His wife had died. He was described as 'a valued friend; a steady consistent, and far-seeing counsellor; one who, so far from being shaken by difficulties, brought fresh vigour to your counsels, whose abilities and eloquence you have had so many opportunities to witness'.[12]

Concurrent with agonised debates by the shareholders, the company was now closely linked to developments on a national level. As a noted engineer, Brunel's championship of the screw propeller was being closely monitored by others. Brunel was to become a major figure in influencing the Royal Navy in its decision to use the propeller.

The Brunel family was well acquainted with the navy. Isambard was born in Portsmouth while his father, Marc, was developing blockmaking machinery for the Royal Navy. Marc continued to act as an occasional advisor to the Admiralty, including conducting some early steamship trials. Until now, Brunel's connections had been indirect, but he was in a different situation since the success of the *Great Western* and his London to Bristol railway. He had real status as an engineer.

It has been suggested that the Admiralty were reluctant to move with the times when it came to steam, especially propeller-driven ships. The Admiralty were early adopters of steam, using paddlewheel tugs and packet boats.[13] Steamships with paddlewheels were also used for amphibious assault and inshore bombardment, but they were vulnerable to gunfire in a front-line engagement. An enemy attack on the paddlewheels could cripple the ship instantly. The screw propeller held out more possibilities for the navy, but it was not a straightforward decision; there were, as Professor Lambert has pointed out, financial, technical, political, tactical and strategic questions that had to be addressed. The Admiralty was constantly bombarded with suggestions from well-meaning and self-interested parties anxious to promote their invention or idea. The Admiralty's strategy was to keep a watchful eye on developments and to get the private sector to spend their money and time rather than rush into costly experiments.[14]

Administratively there were difficulties as responsibility for these new developments were split between two departments. Captain William Symonds was the surveyor of the navy, a dockyard-trained naval architect, but Captain Sir William Edward Parry was the controller of the steam department. Symonds largely ignored Parry and there was no integrated approach to steam warship design. This lack of cohesion was only resolved in March 1850 when the steam department was added to the surveyor's office.[15]

To be effective, the success of a propeller rested both on its design and its position within the ship. Promoting different ideas to the navy were Eriksson, Frances Pettit Smith and others who were eager to make money out of their inventions and get the seal of approval from the Admiralty. While the navy was taking a keen interest in Brunel's activity. Admiral Parry requested that a copy of Brunel's report to the Great Western Steam Ship Company on the propeller versus the paddle wheel might be sent to the First Lord of the Admiralty. By bringing Brunel into the process, the Admiralty secured the best engineering advice from an individual who was independent of the competing designs. Brunel had thus become the leading independent source of advice on the new system through his pioneering work in adapting the *Great Britain* for the screw propeller.[16]

Captain Chappell of the Royal Navy had been appointed to establish trials of the propeller system by testing the *Archimedes* against the Dover steam packet. In his conclusion he said, 'It is evident that in this vessel the propelling power of the screw is equal if not superior to that of the ordinary paddlewheel. Mr Smith's invention may be considered completely successful.' Captain Chappell continued more extensive tests on the *Archimedes*, sailing around the coast of Britain and from Plymouth to Oporto in Portugal.[17]

During 1841 the Great Western Steam Ship Company continued its experimental work with the *Archimedes*, led by Thomas Guppy. The challenge was to fit an effective screw propeller on a large, heavily built vessel. Many experiments were carried out on the *Archimedes*, but the ship had been moved from Bristol before the experiments had been completed. Guppy had a three-arm screw propeller made at the company's works and fitted it on a French Post Office packet ship, *Napoleon*. Built by Normand of Le Havre, this vessel was twice the size of the *Archimedes* and reports of good speeds were obtained.[18]

At the Admiralty, amidst departmental clashes, it was agreed that a new ship would be built due to the importance of the positioning of the screw propeller, rather than trying to convert an existing naval ship. The navy ordered a new wooden vessel similar to the *Polyphemus*, an earlier wooden ship used for propeller trials, to be built at Sheerness. At Brunel's suggestion the Admiralty used Maudslay for all the machinery. The building of *Rattler*, as she was to be named, was closely supervised by Brunel and Pettit Smith and was launched in April 1843, the world's first screw propeller warship. The ship was towed to East India Dock where Maudslay installed the machinery. The first steam was raised in October 1843 and at each stage Brunel ensured the measurement of trials and engine indicators. Well trained by his father, he wanted consistent measurable data. *Rattler* was involved in 32 trials under Brunel's direction for a year between October 1843 and October 1844.[19]

Brunel thus had two propeller trials running concurrently, low-powered naval auxiliary steamers and a high-powered Atlantic liner. These trials led to a reduction in the length of the screw, and longer blades. While the navy adopted a two-bladed hoisting screw, Brunel designed a six-bladed fixed propeller for the *Great Britain*. Throughout this time of experiments Brunel collaborated closely with Francis Pettit Smith, with whom he developed a good working relationship. They used the information from the naval trials to inform their requirements for the *Great Britain*.[20] Brunel's contribution to the successful adoption of screw-propelled warships was significant and in return he gained considerable information from the many naval trials of the propeller, information that he put to good use. It also confirmed him as 'the leading professional engineer of the age'.[21]

Brunel could have made money from patents but he was averse to the clumsy and expensive patents system. It was his view that it rarely 'benefited the genuine inventor, and acted most frequently as a hindrance to the adoption of good ideas and the promotion of further innovation'.[22] Close friends such as Robert Stephenson, Thomas Guppy and Christopher Claxton, and indeed his own father, Marc Brunel, made use of the patents system but Isambard Brunel refused to bend on this issue. This does not mean that he was not open to new ideas; quite the reverse, since he was happy to share his thoughts and worked with others in developing their innovations. Angus Buchanan, who worked extensively through Brunel's letters, suggests that Brunel was quite unperturbed when ideas to which he could lay claim were borrowed or patented by others.

At last, in March 1844, Thomas Guppy could report to the directors that the engines of the *Great Britain* were now completed. 'The investigation of the best forms of a propelling screw, founded on the experimental trials now being made in the government steamer the *Rattler*, by Mr Brunel, has induced him to suggest some slight alterations in the shape of the one we are now making, which is in so forward a state, that within six weeks I expect it will be ready to be tried.' He also went on to point out that as the *Great Britain* was almost complete and the annual maintenance of the *Great Western* had been completed, there would soon be no employment for the foreman and workmen. 'I have, therefore, in pursuance of your instructions, given notice that you will shortly close the works. Their general industrious and steady good conduct demands an acknowledgement from me.' The report referred to the mechanics and other able hands 'who are most of them strangers to Bristol' and there had been hopes that these skills would be retained in in the city. This would have been a big boost for the engineering and iron shipbuilding capacity of Bristol. Elsewhere in the report it had been pointed out that the sale of lease for the works by Mr Ashton, shipbroker and auctioneer, had been so far been unsuccessful.[23]

At last she was ready to come out of her dock. The date of the launch, or rather the floating out, was chosen as 19 July, which was the same day and month as that of the *Great Western*. It was a grand occasion and the Great Western Steam Ship Company had achieved a major publicity coup.

Prince Albert himself came down on the new Great Western Railway to view this new ship and to give it his blessing. The special train for the royal visitor was a celebration of the newly completed railway. On the footplate with the engine driver, J. Kirkham, and urging speed, were Isambard Brunel and his protégé, Daniel Gooch. At the station among the many dignitaries to greet the Prince were directors of the Great Western Steam Ship Company, 'P. W. S. Miles, Henry Bush, Esq., Wear, Pycroft, J. Miles, J. England, J. Guppy, C. Claxton, with R. B. Ward, Esq., and some gentlemen officially connected with the Great Western Steam Ship Company'.[24]

Prince Albert was taken in a procession from Brunel's station at Temple Meads through decorated arches erected in the streets, and thousands of people crowded to see the great event and Prince Albert in person. As the procession wound through the streets, houses had been decorated, including Claxton's residence on St Vincent's Parade, which the exuberant Claxton had covered with white and red rosettes. The Great Western Steam Ship works were decorated with a profusion of laurel boughs. On arrival at the company's site, the Prince was welcomed again and guided to the ship. He stepped on board the ship while she was still in the dock and was conducted over it by Guppy, and after completing an extensive tour internally he then went into the dock to walk the length of the vessel underneath.

After inspecting the ship there was a banquet at which the guests included the Prussian, American and Sardinian ministers and their wives, the Marquess of Exeter, the Earl of Liverpool, the Earl of Hardwicke, Lord Granville, the Marquess of Northampton, the Earl of Lincoln, Lord Charles Wellesley and many more. The long list of names included many of the shareholders, delighted at last to see their new ship in all its pomp and glory being honoured by a special royal visitor. Sir Marc and Lady Brunel were among the 600 guests. The banquet was held in a 'spacious pavilion, formerly the smith's room, neatly fitted up with no traces of its former uses'. The banquet was a deliberate plan to allow time for the water to be let into the dock so the ship could be floated. Meanwhile around the dock and around the harbour there were thousands of people all waiting for this great occasion and ships, boats and barges crowded with spectators waving flags.[25] After the cold buffet and multiple speeches, the Prince and key guests were taken outside to a small pagoda-type building at the entrance to the dock. The dock gates were opened and a small steamer, the *Avon*, took a line so that the ship slowly emerged out into the harbour.[26]

> As soon as the immense mass of iron floated out, a bottle of champagne, destined for the christening, was placed in the hands of Mrs Miles, which she tended to the Prince; His Royal Highness, however, graciously expressed his wish that the lady should perform that part of the ceremony itself; accordingly she dashed the much honoured bottle against the side of the vessel while the Prince pronounce the name and it floated forth a vessel built by British men, with British men on board, and named the *Great Britain*.[27]

All was well. The ship was afloat and could be seen in all her magnificence. Prince Albert left to return to London via the Great Western Railway.[28]

Once floated out into the harbour, the ship, now 40 feet longer than originally proposed, remained there for the next year as she was fitted out. The final design of the propeller had not yet been agreed as they were still waiting for the experiments to be completed on the *Rattler*. In the event the propeller had to be designed and fitted without that final data. The big problem now was getting the ship out of the harbour and up to London. Despite previous difficulties, the company had been optimistic that the Dock Company would be prepared to widen the walls of the lock from the floating harbour into the river, but this was a major task and an expensive one. In their confidence they advertised the sailing times from Liverpool.

> *Great Western* and *Great Britain* Steamships. – it will be seen from an advertisement elsewhere, that the *Great Western* is appointed to sail from Liverpool, for New York, on the 27th of April next, and that the *Great Britain* is also appointed to sail from Liverpool for the same destination, in May. Captain Hosken has succeeded to the command of the *Great Britain*, and Captain Mathews, late principal officer on board the *Great Western*, has been appointed in his place.[29]

While the ship was moored in the floating harbour, she continued to attract a stream of visitors, including foreign nobility such as the Duke of Bordeaux. On his sightseeing tour of the West Country, which took in treasures of the great art collector, William Beckford, he and his party visited Bath, viewing Victoria Park, the Assembly Rooms, the Pump Room and Baths, and Prior Park. The next day they headed for Bristol to the Great Western Cotton Works and then examined the *Great Britain* steamship.[30]

The situation was looking better for the benighted proprietors. By March 1844 Henry Bush as chairman could report better financial news to them. The economic situation in the United States was improving and operating out of Liverpool continued to be advantageous. In 1843 the *Great Western* had earned £33,406 and expenditure had been £25,573; the comparison in 1842 was an income of £30,830 and expenditure of £28,615. The main subject under discussion was the continuing difficulties with the Bristol Dock Company. Negotiations were slow and the directors of the steamship company had engaged Brunel as their engineer to widen the dock. 'Mr Brunel's services being engaged for the company, to arrange the passage of the *Great Britain* into and out of Cumberland Basin, whether by any plans of his own, or by negotiation with the directors of the dock company'. Accordingly, he had a meeting with them on 6 February and subsequently sent a letter outlining his proposals.[31] Optimism was not met with success as the Dock Company stubbornly refused to accept any liability and suggested the matter would need to go through an Act of Parliament. For six months the arguments went backwards and forwards until an agreement was finally reached.

By December there was another extraordinary meeting of the shareholders:

A meeting of the Great Western Steam Ship Company has been held to empower the directors to raise additional capital, by the creation of new shares, to defray the expenses already incurred, and to complete the equipment of the *Great Britain* for sea. The necessary power was granted, and there is every probability that the *Great Britain* will be able to make her first voyage in the spring. The machinery of the *Great Britain* was set in motion one day last week, for the purpose of trying the screw, which was found to act in every respect to the satisfaction of the engineer.[32]

At last, on 10 December, with matters finally resolved with the Dock Company, the ship was ready to move and the tides were right for her to go through the lower lock into the River Avon. It was almost a total disaster. Captain Claxton was on board the steam tug that had the *Great Britain* under tow. He knew the margin for error was slight and as he watched he could see that the ship was touching the lock walls on either side. The ship jammed itself in the lock gates and, acting with commendable speed, he gave orders for the ship to be hauled back. This was essential as the tide was already beginning to fall and, if they had not reacted at that moment, the ship would have been firmly wedged in the open dock entrance. No time could be lost to resolve the issue because the tides had passed the highest point. Brunel swung into action, leading the working party to make the alterations.[33] He was due at a meeting of the directors of the South Wales Railway and wrote to apologise for not attending:

We have had unexpected difficulty with the *Great Britain* this morning. She stuck in the lock. We did get her back. I have been hard at work all day altering the masonry of the lock. Tonight, our last tide, we have succeeded in getting through but being dark we have been obliged to ground outside, and I confess I cannot leave her till I see her afloat again and all clear of our difficulty here. I have as you will admit much at stake here and I am too anxious to leave her.[34]

The necessary alterations were made to slightly widen the entrance again so that on 12 December 1844, 18 months after she left her dry dock, at long last, she was towed down the River Avon and as she went the boilers were filled, the fires were lit, and steam was raised.

An anonymous Bristol poet wrote in celebration of the occasion:

Six masts – like princely sons to bear!
Great Britain for my name
My smoke trail black on the sun bright air
My screw as swift, and my sails as fair
As the trumpet voice of fame!

Following the first crossing by the *Great Western*, Claxton had published a pamphlet celebrating the success with the engineer's log, a passenger diary and details of the ship's build. He now did the same thing for the *Great Britain* in advance of the ship going into service. It was typically bullish in its statements and descriptions and looked to secure this ship's place in maritime history.

This splendid iron ship – the largest vessel we believe in the world – was launched, or rather floated off, from the dock at Bristol in which she was built, on 19 July, 1843 in the immediate presence of Prince Albert, the large concourse of noblemen and gentlemen, the families of the first distinction from nearly every quarter of the kingdom, as well as many thousands of spectators belonging to that town, and congregated on the adjacent Heights, and every available point of view on shore, or from vessels on the river. The untoward delays that afterwards arose in getting the vessel ready for sea, are already before the public. Every difficulty has happily been overcome; and as the vessel has already most satisfactorily solved the problem involved in the magnitude of her construction, and her peculiar motor propulsion (from which a new era in ocean steam navigation will henceforth be dated), we have taken steps to gratify our readers by a more detailed account of the Leviathan and her machinery, with statistics, illustrated by woodcuts, presenting views of a whole, machinery, et cetera so that an accurate idea of the whole may be attained at a glance.[35]

His words on the ship were that 'her model is somewhat peculiar, yet according to the taste (when she was built) of many nautical men and the speech has since attained together with her good sea qualities prove that their opinions were well founded.'[36] Claxton's use of the word 'peculiar' was the Victorian term meaning individualistic rather than odd. The rig he described as a six-masted schooner with fore and aft sails and lugger top sails. She could carry a total area of about 16,000 square feet of sail, but more masts meant smaller sails, which required fewer men to handle.[37] 'Nothing is so different to handle under a variety of circumstances as the sails of the steamer unless the engines be stopped, which should never be allowed in Atlantic steaming, where onwards and for ever onwards is the rule.'
In those rare circumstances where the ship progressed on canvas alone, the problem was that the big screw would slow her down. However, in an emergency the captain could disconnect the screw (with some difficulty) and it would then just revolve on its own, causing less drag.[38]

The rigging was wire and the *Great Britain* was the first large ship to be fitted in this way. The advantage was that it was lighter than rope strength to strength, although this was subsequently modified in the light of experience. The emphasis on the innovation of the use of iron and a propeller sometimes obscures the fact that the *Great Britain* was the world's first six-masted ship. The six-masted schooner did not make an appearance until the early twentieth century.[39]

The *Great Britain* had four decks and the ample spa deck was 308 feet in length. 'The engines are somewhat on the patents of Sir Mark Brunel, with the cylinders, in place of being upright, standing on an angle of about 60. The main shaft of the turning of the screen, and which is of a great length and large diameter, was made at the Mersey Ironworks; and is itself a great curiosity.' The decks below were lit with eight skylights for the fore saloon and one large light over the engine room. 'The under decks and apartments have borrowed lights on these and also circular lights on the sides of the ship, the latter of plate glass an inch in thickness.'[40] The ship carried iron lifeboats, the patent of which belonged to Guppy. There were four of these boats, each with iron buoyancy tanks, and they were 30 feet long, 8 feet broad and 5 feet deep. Additionally, there were two boats of wood in davits and one large lifeboat on deck. Together they were 'capable of carrying 400 people'.[41]

The ship was beautifully decorated externally. The hull was painted black but she had a white stripe painted with imitation black gun ports. 'The bowsprit is proportionately short owing to the great length of the vessel. The bow is enriched with carved work: in the centre the Royal Arms, surrounded by emblems of the arts and sciences of the Empire, and (in illustration of the power and speed of the ship) representations of the thunderbolt of Jove and the caduceus of Mercury.'[42] The Royal Coat of Arms were supported by a gilt lion and a unicorn. Also displayed were various symbols of Victorian trade, art and industry, including a carpenter's square, gear wheels, a rope coil, a sheaf of corn, a globe, a lyre and trumpets, an artist's palette, a book and a bunch of flowers. On the stern was a unicorn supporting the arms of the City of Bristol.[43]

There were '26 state rooms with one bed each and 113 with two, so that in addition to her crew, officers, fireman et cetera she can accommodate 252 passengers, each of whom can be provided with a single bed and that without making up a single sofa or any other temporary convenience'.[44] This was double the number of passengers carried by the *Great Western*. In its first ship the company had advertised just one class, superior class, but with the second vessel there was a distinction between first class and second class. First class was at the stern of the ship with a dining saloon with cabins around it on the lower deck. Above it was the promenade saloon, again surrounded by cabins. There was a similar arrangement for second class in the forward part of the ship.[45] There were water closets in both areas and a lady's boudoir. Stewardesses were also accommodated close by. The lady's boudoir had a settee and a table. Family cabins were as yet not standard, but there were two cabins that could be arranged as one space with communicating doors.[46]

Ever conscious of the importance of accommodating the ladies on board, Claxton specifically pointed out that about 12 of the cabins 'on each side of the deck will be reserved for the ladies as they are made to communicate with two commodious lady's boudoirs. The advantages of this arrangement must be obvious as ladies who may be indisposed or in negligée will be able to reach their sleeping berths without there being the slightest necessity for their appearing in public.'[47] The captain's accommodation was close to the first-class passengers.[48] The cabins were much the same as on the *Great Western*, 6 feet wide.

The main dining saloon was 98 feet 6 inches long by 30 feet wide, a 'really beautiful room'. Claxton described the decoration: 'The walls of the after or principal promenade saloon are painted in delicate tints and along the sides are several fixed chairs of oak.' As before, the decoration of the ship was carried out with an eye to good taste. The well-known firm of Frederick Crace was again commissioned. The family firm of Crace had been established in 1750 and was a firm favourite with the royal family and leading architects of the day. Frederick Crace worked almost exclusively for the royal family and was responsible for the decoration of the Prince Regent's Royal Pavilion at Brighton.[49] On the ship, the decorators had done their work well, with columns of white and gold, pilasters in the saloon painted with oriental birds and flowers, and carved and gilded archways to some doors. All the saloons were covered in Brussels carpet purpose-made by Messrs Mogg of Bristol.

The saloon was fitted with rows of dining room tables so that 360 people could sit down to dinner at one time 'with perfect convenience and comfort'.[50] The tables were laid out with three long tables and at full capacity each diner would have about 17 inches. As has been pointed out, diners would certainly have had to eat very politely with elbows well tucked in and the portlier members among the passengers would not have been very popular.[51] Passenger fitness was not forgotten: 'Eight walks round the principal deck are about equal to a mile in length.'[52]

For Claxton the most interesting portion of the whole structure 'is the machinery, and the screw, by which she is propelled. The latter is on the same principle, but slightly modified, as that invented by Mr F. P. Smith of the Patent Ship Propeller Co (who supplied it) and who some years ago exhibited it in the *Archimedes*. There were plenty of detailed illustrations of all aspects of the ship and its machinery.'[53] He paid fulsome tribute to Pettit Smith and 'that small body of gentlemen who built the *Archimedes*'.[54]

The problem with the *Archimedes*' screw was that the machinery for transferring power was very noisy, indeed the noise was described as intolerable. This would not do on a luxury passenger ship. Claxton credits Brunel as introducing leather straps in the *Rattler* to avoid the noise. The noise was caused by cogwheels but in the *Great Britain* they used chains, 'for the same object, although weighing seven tons, (which) worked without the slightest noise'.[55]

Several trials were carried out locally to test the engines and propeller. The first was on 8 January and there was quite a group on board, a 'numerous party of proprietors, and several engineers and scientific men'. Unfortunately, there was dense fog and despite waiting several hours for it to lift the pilot could only be persuaded to travel a short distance, mainly to please the visitors.[56]

Another trial was a run down the Bristol Channel to Ilfracombe and back. Captain Hosken was in charge having transferred from the *Great Western*. Ewan Corlett wrote, 'Indeed *Great Britain*'s performance on trials was remarkable. There was little or no difficulty with her machinery and no need for alterations, the ship met her intended speed, she steered well; and everything was in order in spite of the novelty of the construction of her and of her means of propulsion.' He went on to say, 'Only naval architects and ship builders can appreciate the enormous achievement this represented in the technological context of the day.'[57]

On 14 January 1845 the ship was entered into the ship registers at the Custom House in Bristol and received her unique number, 25967. As was required under the deed of settlement, she was registered to John William Miles and Thomas Bonville Were, trustees of the Great Western Steam Ship Company, who held it on behalf of the proprietors.

After these local trials on 23 January 1845 the *Great Britain* sailed for London. There had been much curiosity about how the propeller-driven ship would manage in heavy seas and she was tested immediately in a heavy cross sea as she left the shelter of Avonmouth. In gale force winds off Lundy Island she was 'struck on the starboard bow by a tremendous sea, which must have contained two or three thousand tons of water. The shock for a moment seemed to paralyse the vessel, and to bring her to a standstill; and even to make the engines themselves jerk in such a manner that, had they not been fixed to a nicety, might have done them some damage.' The ship recovered but 'three of her starboard bow bull's eyes were stove in, together with their frames; the diagonal bands of her forecastle deck were bent, the woodwork started two inches upwards and a portion of her carved figure-head carried away, and the wooden fittings of her bulkhead and the iron sheathing of both bows split above deck in two places.'[58]

Guppy, in a later report, referred to the 'inefficiency of the stokers' so that 'steady steam pressure was not maintained'. It was a very uncomfortable voyage for all those on board, particularly the engineers and firemen.[59]

The crew of sailors consisted chiefly of that indifferent class usually shipped on short runs, to whom of course the rig of the ship was perfectly new. Some of the engineers stood well to their duty, but others, and nearly all the stokers, were completely knocked up with seasickness. The deck was encumbered with at least 30 tons to 40 tons of chain cables and materials; and the coal stored chiefly in the upper bunkers for the great convenience of working it, with so few men. Consequently, with no weight in her bottom, the centre of gravity was raised so high, that the rolling which was considerable, but very easy, is not surprising.[60]

The vessel continued on her way to London causing a sensation when she was sighted, her six masts giving her 'an extraordinary appearance'.[61] The *Illustrated London News* reported, 'As she passed up the river the crews of every vessel ran on deck to obtain a view of her. Her extraordinary length and her singular appearance rendered her an object of considerable attraction.'[62]

The length of the ship was a wonder and as she arrived to take up her moorings at Blackwall, she swung round with the tide and briefly became a bridge between the two sides of the river, displaying her length. 'Some idea of her extraordinary length may be formed when it is stated that she is upwards of one hundred feet longer than either of our first-rate line-of-battle ships, the *Queen, Caledonia*, and *St. Vincent.*' The critics were confounded and it was reported that 'it is acknowledged in all quarters that the complete success of the screw propeller has been triumphantly established and it is confidently stated that the *Great Britain* will attain a speed of 14 knots, or more than 16 miles an hour.'[63]

One sad event disturbed the general celebrations when a stoker from the ship drowned off Blackwall. He and five others had been ashore for the evening on 17 March and on returning they hired a waterman to take them back. He pulled up alongside a barge by the side of the ship, but in the dark, and with a strong tide running, his boat was caught under a chain and swamped. Moses Porter was drowned, but the rest, who were clinging to chains, were rescued by their fellow crew of the *Great Britain*. The coroner's jury found that they were all sober at the time and declared a verdict of accidental death.[64]

The ship remained in the Thames and, as her predecessor had, became a great attraction. Until her marriage Queen Victoria had shown little interest in technological innovations and the *Great Western* had been untroubled by a royal visit. In 1844 things were quite different, Prince Albert was a great supporter of new technology and had helped to launch the ship, so he and the Queen made a visit to examine her on 25 April.

It was a visit in great style. They left Buckingham Palace in an open carriage and four, escorted by a party of light dragoons and proceeded to Greenwich. Here the royal party embarked on board HMS *Dwarf*, commanded by Lieutenant Beauchamp Proctor.[65] The party included the Board of Admiralty and the Queen's progress from Greenwich Palace was preceded by a state barge with the Lord Mayor of London and many other steamers packed with spectators following the procession. It was a fine day and thousands crowded the shores. Once on board the *Great Britain*, the captain, who was referred to by his naval rank of Lieutenant Hosken, was presented to her Majesty by Lord Hawarden.[66] The Queen was met on board by the chairman of the steamship company and directors including Were, Miles, Pycroft, Claxton and Guppy. Brunel and Pettit Smith were also in attendance. The party was then conducted through the vessel. After the tour of the deck to see the sheer scale of the ship the visitors went below to inspect saloons and state rooms.[67] The Queen duly expressed amazement at the enormous length of the ship, 'which was one-third longer than any line of-battle ship in the service'.[68] They were then shown a model of the engines and screw, Isambard Brunel himself explaining how they worked. Guppy showed them around the engine room itself, where the Queen was particularly fascinated by the chain that turned the screw shaft. The Queen was shown three different models of propellers and presented with a gold model of a propeller by Pettit Smith. Captain Claxton gave her copies of his description of the ship.[69]

On leaving she addressed Captain Hosken, and said, 'I am very much gratified with the sight of your magnificent ship, and wish you every possible success in your voyages across the Atlantic.' The Royal party then returned to Buckingham Palace. 'Her Majesty wore a tartan silk dress, and Paisley shawl, with a light blue bonnet, and appeared in the enjoyment of excellent health.'[70] This was contemporary code as she was six months pregnant with her fourth child at the time.

Foreign nobility lost no time in following the queen's example. Her Royal Highness Stephanie, Grand Duchess of Baden, was accompanied by the Baroness de Strumfeder 'and her usual retinue'. Conveyed in the steam yacht HMS *Dwarf* from Brunswick Pier, the party was escorted on board the *Great Britain* by the Chief

Officer, Mr Hedger, standing in for Captain Hosken. Mr Harman, the engineer, explained the machinery to Her Royal Highness and her party and the Duchess was sufficiently curious that 'she even went through the narrow passage in which the screw shaft works when the vessel is in motion'. This must have been interesting in view of the voluminous skirts adopted by most fashionable ladies of the time.[71]

The ship then entered East India Dock where the public was able to visit, and the *Great Britain* remained for a whole five months in the Thames before at last heading to Liverpool.[72] Thousands of people waved the ship off from the Thames as she headed into the English Channel to begin her commercial role, with 80 passengers on board.

3

'MONARCH OF THE OCEAN': THE *GREAT BRITAIN* GOES TO SEA

Even after such a lengthy stay in the Thames, the company was in no hurry to get her to Liverpool. They were determined on a stately promotional progress and to show her off where they could. The plan was to call in at several ports on the way, but not all ports were welcoming. In Southampton an enquiry had been made asking if the Southampton Dock Company would deepen the entrance into the tidal basin to admit the *Great Britain;* there was a suggestion of a link with an excursion train from London. But as the newspaper reported, 'The Southampton public will regret, though they will not be surprised, to hear that they have been deprived of an opportunity of viewing the *Great Britain* steamship in their port, by the strange obtusity of the Director of the Directory. The reply was as rude as concise – he should do no such thing.'[1]

So the ship steamed past Southampton and called at Plymouth where Captain Claxton had already visited the previous month to arrange a suitable berth for her, a mooring at 'Mr Gill's pier'.[2] Spectators thronged both onshore and in small boats in the harbour. About 14,000 people visited her while she was in Plymouth and she also made a day trip to the Eddystone Lighthouse on 20 June carrying 600 passengers.[3] She continued onwards making stately progress and arrived off Falmouth early that evening. The next day she headed for Kingstown, the port for Dublin, to another enthusiastic welcome and was joined by new passengers. A group from Liverpool boarded the *Iron Duke* and set off to Kingstown for the sole purpose of meeting the *Great Britain* and returning in her to Liverpool. As the *Great Britain* left Kingstown there were many spectators cheering and waving her off.

On board at 3 o'clock that afternoon dinner was announced and the reporter from the *Liverpool Mail* estimated at least 50 gentlemen sat down to a sumptuous meal in the dining saloon. This space he described as a lofty and spacious apartment with 'so many pillars supporting it that one might fancy that he dwelt in marble halls'. Captain Claxton was on board and presided over the meal and he led the inevitable toasts with one to the assembled company. On this occasion Captain Matthews' health was drunk and he replied in 'flattering terms' about the *Great Britain*. Then Captain Hosken's health was toasted followed by that of Mr Guppy the engineer. Guppy referred to some of the difficulties that had been involved in the construction but referring to the size of the ship said he 'only wished they would give him an order for one a deal larger'. Claxton proposed the health of Mr Brunel who was not on board but whom he said 'was a deal better known at Bristol than he was here'. Then came the toast to the navy. Mr Miles from the Great Western Steam Ship Company was on board and Mr Bright, and the Liverpool agents, Gibbs, Bright and Co., were also toasted.[4] Finally Claxton announced that he now needed to go on deck and, at last, the *Great Britain* came into Liverpool escorted by 'steam boats, sailing vessels, and yachts of every calibre, out viewing one another in doing honour to our gallant vessel'. The bell at the lighthouse rang and guns were fired from the shore to notify the arrival and the ship responded in style. The correspondent particularly singled out Captain Hosken, who was very well known from his time as commander of the *Great Western*.[5]

As to her skilful and intrepid commander, he is so well known to the Liverpool public, that any praise would be superfluous. Still we must say that his gentlemanly, kind, and courteous bearing won for him the respect and esteem of all on board.

In him the company have an old and tried friend, and the flattering reception he experienced must have convinced him of the high opinion our townsmen entertain both of himself and of his noble ship. The remembrance of this passage, we have no doubt, will long remain in the minds of many of those who had the pleasure of sailing with him on this occasion. The only drawback to our pleasure was the total absence of the fair sex, not one of whom honoured us with their presence; but whether this was caused by the wind and rain, we know not; but if they had known what a nice lot of gentlemen were on board, we are sure neither one nor other would have kept them.[6]

The news correspondent finished by listing the large numbers of people that had visited the ship at the various ports and they felt that 200,000 was not far from the mark as an estimate.[7] At last, the great ship was docked in Liverpool and was open to the public. Her arrival there had long been anticipated.[8]

The Great Britain Steam Ship
The Great Britain steamship, Coburg Dock will be
OPEN for the INSPECTION of the PUBLIC until
Wednesday 23rd inst.
Admission every day except Sundays and Friday, the 18th
from 8 am to 9 pm
at 1s. Children 6d
Engine Room, 6d additional
By order of the Directors, GIBBS, BRIGHT and co
North John-Street, Liverpool, 11 July 1845
THE GREAT BRITAIN STEAM-SHIP.
TO be HAD of the TICKET TAKERS, on board the Great
Britain, a PAMPHLET, giving a full description of that Ship,
with a Drawing and Plans of her accommodations, and
also an account of her voyage to LONDON. By Captain
Claxton, R.N., a Director of the Great Western steam-ship
Company.
PRICE SIXPENCE.[9]

It was estimated that 2,500 people a day visited the ship.[10] The directors of the Great Western Steam Ship Company donated £100 to the following charities: Northern hospital £25, Southern hospital £25, sailors home £20, dispensaries £20 and £5 each to the floating church and the floating chapel. The total number who went on board the *Great Britain* during her time was reported as in excess of 38,000 persons, which makes the charity donations seem rather mean.[11]

But she was a very expensive ship. The final cost of the *Great Britain* was about £120,000 and running costs were high. The annual cost of the crew has been estimated by Ewan Corlett. For the seamen, firemen and stewards he estimated £7,300 and including the officers he totalled about £9,000. Looking at this wage bill and the cost of coal and income from fares and cargo, he supposed that the directors might have anticipated 34 per cent return on capital 'enough to justify the entire venture. These figures, though to some extent speculative, do indicate that the *Great Britain* was no white elephant; given luck and continuing service, she could have been a profitable venture for the Great Western Steam Ship Company.'[12]

Just as with the *Great Western*, the *Great Britain* attracted the sceptics. In an article in the *Nautical Magazine* in 1841, a writer with the pseudonym *Nauticus* queried the choice of screw propulsion and suggested it was a 'rather hazardous thing on the part of the company, to adopt an invention, the success of which they are not as yet fully satisfied... One voyage however from Bristol to New York, will be quite sufficient to decide this question.'[13]

New York had been expecting the new ship for some time and perhaps the delays in the whole project led to the following rather caustic remark:

The *Great Britain* steamship will leave Liverpool on 25 July for New York. Those who expect to see a beauty of a boat, or much fine carving and gilding, will be disappointed. She is built for use, not show, and her only claims on attention are, her stupendous proportions, and her qualities as a sea boat.[14]

Hosken as the captain of the new ship had already travelled to New York to make arrangements for berthing the *Great Britain*. Having commanded the *Great Western* for so long he was well known in that city, having made many passages on her. Hosken was at the height of his career. An ex-royal navy officer, he was assiduous in his treatment of the many wealthy passengers he had carried on board. As a result, he had a large address book of the great and good and his public profile abroad was higher than anyone else in the company. He was described by one wealthy New York lady as 'the best, kindest most gentlemanly of all captains'.[15]

On the afternoon of 26 July 1845 at 3:20 pm the *Great Britain* left her moorings and headed for her first crossing of the Atlantic. The number of passengers on this maiden voyage was disappointing, just 45 people and 360 tons of cargo.[16] The *Great Britain* fares began at 35 guineas, then 28, 25, 22 and the lowest fare was 20 guineas. The 18 English passengers were outnumbered by 22 from the United States. As with the first passage of the *Great Western*, the majority of passengers were male, just two women travelled out. One was an American travelling with her husband and there was also one Englishwoman travelling alone, Anne Vyce Palin, age 36. There was the usual mixture of merchants and military men. On board the ship was a man called Wood who gave his occupation as confectioner, as did an F (sic) Suchard, age 45, a Swiss national.[17] This is quite possibly Philippe Suchard, who was just gaining success as a chocolatier, and there are two connections here. First, he was familiar with the United States having visited in 1824 and subsequently written of his journeys. Second, he was also interested in marine steam, having introduced the first steam boat, *Industriel*, on Lake Neuchatel in 1834.[18]

Among the military men was Captain William Morris of the Royal Navy who returned with the *Great Britain* on the very next passage, which rather implies that his main interest was in the ship itself rather than travelling to the United States. The Royal Navy was still carrying out trials on the propeller, so observations from Morris of its performance across the Atlantic were bound to be of interest to the Admiralty. There was poor weather on the voyage with westerly gales all the way and later thick fog, but overall it was an uneventful passage. Her engines worked well and were not stopped until her arrival off Sandy Hook.

The passengers, like their predecessors on the *Great Western*, were fully aware of the importance of this first passage across the Atlantic in an iron screw-propelled vessel and there was the usual organisation by passengers to establish a committee to 'draw up an address to Captain Hosken, expressive of their sentiments of the performance of the steamship *Great Britain* on her passage from Liverpool to New York, and of her adaptation to the comfort, safety and convenience of passengers'. They formed a committee chaired by some distinguished passengers. The chairman, Lieutenant Colonel Everett, was a Fellow of the Royal Society, an organisation of which Brunel had long been a member. Other committee members included two United States Navy officers, Elisha Kent Kane age 36 (a surgeon in the navy and later to be known as a famous explorer, participating in two Arctic expeditions to rescue Sir John Franklin) and Lieutenant H. A. Wise, plus Captain Morris of the Royal Navy and a retired British Army officer, Septimus Crookes. Conscious of history, and no doubt of good publicity, a letter was composed to be signed by all passengers on board.

Steamship *Great Britain* 7 August 1845
To Lieutenant James Hosken RN
Commander of the steamship *Great Britain*
Dear Sir
We the undersigned passengers in the *Great Britain* steamship, having accomplished our voyage to our entire satisfaction, present yourself and the company whom you represent our congratulations upon the successful result of this, the first practical attempt to cross the Atlantic in a vessel propelled by the Archimedean screw propeller.

The considerations which especially led us to this step are based upon the magnitude of the *Great Britain* and the nature of her materials, which, taken in conjunction with the character and machinery and the novelty of its application, gave rise to an excited state of public opinion, which attached the highest experimental importance to the successful termination of our passage.

Our opinion derives an additional value from the fact not only of having successfully faced adverse winds and heavy sea but that during this period of four days duration the operation of her machinery never experienced the slightest interruption.

We feel especially called upon to allude to the fact, as interesting to the admirer of the vessel, as important to our own comfort, under the influence of an ordinary breeze, there is toward the head of the vessel absolutely no vibration whatever caused by the machinery; that the vibration at the engine and inward the central part is reduced to a mere tremulous motion; and that even toward the stern where the greatest effect might be expected, it is far less than is usually experienced in vessels propelled by paddle wheels.

In concluding this expression of our satisfaction, we simply confine ourselves to congratulations upon an experiment, in which you have taken such an effective and personal interest, deeming it unnecessary, in assuring you of our regard, to add our commendations of those high qualities for which you are already so well known and appreciated.

We have also to express our entire satisfaction with the luxuriant supply of the table and the excellent arrangement of the stewards' department
We are dear Sir your sincere well-wishers
(signed by all the passengers)
steamship *Great Britain* August 9, 1845[19]

This was duly reported in the New York papers, whose attitude changed considerably when the ship arrived. 'Arrival of the Monarch of the Ocean. She is truly beautiful. She is unquestionably the largest and most magnificent specimen of naval architecture that ever floated' enthused the newspapers and thousands crowded the shores to catch a glimpse as she arrived on 10 August. Her triumphant arrival was reported not just in the United States papers but also widely copied back in Britain and Ireland.

There were some of our citizens congregated at the Battery yesterday afternoon, to see the magnificent steamer *Great Britain*; and she was well worth seeing. She moved very slowly through the water, which was scarcely disturbed by her propeller, and as she passed the wharves, the sailors looked in astonishment at the six-masted monster – for monster she really is. She lies at the Tobacco Inspection Wharf, and I presume will be visited by thousands of our citizens.[20]

The *New York Herald* called her 'The monster of the deep, a sort of mastodon of this age'.[21] Her arrival on a Sunday afternoon enabled thousands to crowd the vantage points to watch.

This magnificent steamer, under the skilful management of Captain Hosken, came up the bay in beautiful style, after her passage of 15 days over the Atlantic. She was gaily decorated with flags, and it was not a little singular that few or none saw the stars and stripes. This excited some surprise. Was the American among the flags displayed! The great problem, whether or not a steamer of the magnitude and construction of the *Great Britain*, and with her principle of propulsion, could make a successful trip across the ocean, is now satisfactorily and happily solved.[22]

The statistics were reported by the newspapers. She had arrived at Sandy Hook at 1 p.m., 10th of August, and made the passage in 14 days and 21 hours. The whole distance from Liverpool to New York was 3,304 miles. The *Great Britain* steamed this distance, against foul winds and cross seas, in 14½ days, or 354 hours, giving an average speed for the whole voyage of somewhat more than 9½ knots or nautical miles (which was equal to 10¾ statute miles) per hour.[23]

From Boston came the reports of 'this monster vessel safely arrived in the Bay of New York' and speed comparisons were made with the mail ships subsidised by the British Government.

From data in possession of Her Majesty's Government, it can be proved that neither the Halifax nor the West India steam-packets have been able to maintain in their outward passages (say for period of six months) an average speed equal to what has been achieved by the *Great Britain* on her first voyage to New York.[24]

Some were not impressed with the new arrival and preferred the existing steamers run by Cunard.

> The general impression among the Americans, says a correspondent of *The Times*, is that she is not as large as they expected. To be sure, she is not as long as some of the North River boats, some of which are upwards of 350 feet in length, 16 of them going to a mile; but then, they would not do for ocean steaming, and in point of breadth are far below the *Great Britain*. Her model is not greatly admired, and they call it clumsy. On the whole, she has not made such a sensation as perhaps her proprietors expected, and is not considered here as a safe boat. People in America will prefer the Boston steamers or the packet-ships.[25]

The reporting of the lack of the American flag was taken up by other papers, who corrected the reports and described the ship on arrival as 'gaily dressed with colours, which streamed from each of her masts. At the gaff the large union flag of England floated gaily in the breeze; the first mast had the Austrian flag; the second, the Russian; the third, the Spanish; the fourth, the French tricolour; the main, the Union Jack; at the foremast, a blended flag of England and America, the stars of the latter country blending with the blue, white, and red, the Union of England, and at the lower quartering, the stripes.' This blending of the national flags, intended as tribute, led one wit to say 'I'm blowed if Great Britain hasn't annexed we!'[26]

Just like her predecessor the ship was a major attraction, visited by thousands who flocked to see this new wonder. The ship was opened to the public at 20 cents a head, and 12 cents extra to see the engine-room. The crowds 'all expressed themselves being greatly delighted and surprised at her internal arrangements'. Captain Hosken was feted and followed by crowds wherever he showed himself. Visitors came in from Boston, Baltimore. Philadelphia and elsewhere to see the big ship for themselves. The *New York Daily Tribune* commented: 'We admired the plain and solid style adopted in all parts of the *Great Britain*, so simple, so judicious, so easily kept clean, so truly English!'[27]

Captain Hosken entertained leading citizens on board the ship. The *Tribune* newspaper noted:

The visitors, who comprise most of the leading merchants of the city, Commodore Jones, Colonel Bankhead and numerous other visitors from the city and Governors Island et cetera all expressed themselves highly gratified with the admirable arrangements of the noble vessel. After the inspection of the vessel, the company sat down to an elegant dessert, and various toasts were given, which we are reluctantly compelled to omit for want of room. The company separated highly pleased with their inspection of the ship and their hospitable entertainment.[28]

Ironically, in the same addition of the *Tribune* there was an advertisement for a book of lectures by Dr Dionysius Lardner.

The ninth number of Lardner's lectures will be ready to deliver to book sellers and agents this day 28 August. It contains lectures on protection from lightning, magnetism, electro magnetism, the thermometer, atmospheric electricity and evaporation. The back numbers may still be obtained price $0.25 each. The work will be completed in 14 numbers.

It was he, as a noted scientist in Britain, who had loudly headed the strong criticism regarding the feasibility of steam across the Atlantic and was now, following a marital scandal, earning his living by lectures in New York.[29]

The ship remained in New York for 19 days and an estimated 21,000 people visited her. An official dinner was given on the last evening for Captain Hosken at the grandest hotel in the city, Astor House.

New York 26 August 1845
to Captain Hosken of the British steamer *Great Britain*
Dear Sir, the undersigned, a committee on behalf of many of our fellow citizens, who feel desirous to express the gratification they have experienced by the visit to our port

of the noble specimen of naval architecture under your command, and who also testify their sense of the merits of her commander, beg leave to invite you to dine at the Astor House, on Friday evening, at 8 o'clock, in order that in a friendly rather than a formal way they may assure you how highly they appreciate the enterprise of your company, how fully they prize the taste and the skill so conspicuously displayed, and at the same time, being able to tender their best wishes for the success and prosperity of the *Great Britain* and her commander
your friends and obedient servants
Samuel Jones, Philip Hone, Jas. D'P Ogden, François Griffin, Wm S Miller, DC Colden, FH Delano.[30]

These men were the New York establishment; Chief Justice Samuel Jones, the diarist Philip Hone, James De Peyster Ogden, a merchant, President of Brooklyn harbour and the Nautilus Insurance Company, William Starr Miller, Congressman for New York and F. H. Delano, a wealthy New York merchant. Chief Justice Jones presided, assisted by James De Payster Ogden. The dinner guests in typical style filled their glasses to drink a series of toasts:

The merchants of Bristol – the first to risk their wealth to transatlantic steam navigation. The thanks of both nations are justly their due.
The President of the United States
The Queen of Great Britain and Ireland
Captain Hosken, whose skill and deportment have secured the confidence of the public with a well-deserved popularity. May the measure of success correspond with the magnitude of his command.
The memories of Watt and Fulton – in the *Great Britain* we witness the grandest triumph of their art, and the proudest moment of their genius.
The Pacific influence of steam – it makes all nations neighbours, and neighbours should never quarrel.
The cities of Liverpool and New York – Honourable competitors in commercial enterprise.[31]

It is noticeable that Isambard Kingdom Brunel, the mastermind behind the ship, whose father, Sir Marc Brunel, had been a citizen and engineer in New York, did not get a mention anywhere in these toasts.

Despite the significant publicity the bookings for the return passage were low. The ship returned to Liverpool with 53 passengers, 1,200 bales of cotton and other merchandise. There were 12 women passengers onboard and the English passengers were considerably outnumbered.[32] There were difficulties on this return passage with the engines not giving full power, the loss of a main topmast which snapped, and although the lack of vibration was praised by the passengers, the same could not be said for the tendency of the ship to roll from side to side.[33] This tendency to roll did not help the firemen. If the firemen could not supply a steady and constant supply of coal the engines would of course lack power.[34] The snapping of the main topmast was due to 'there being insufficient seamen to take in the topsail in a sudden squall of wind'. A significant number of the original crew had deserted in New York just before sailing and Hosken had problems recruiting sufficient to replace them at such short notice.[35]

On her next voyage in September 1845 desertion in New York was still a problem and the first crew agreement to survive is for this voyage. In total 126 crew were employed, 55 had remained with the *Great Britain* from the previous voyage and the rest were from other ships, including a trimmer from the *Great Western*. Crew were not continuously employed and they signed the crew agreement for each journey, in this case from Liverpool to New York and back to Liverpool. They could then sign up with any vessel for the next voyage. In New York 14 men deserted and the majority were seamen; six able seamen, four ordinary seamen, one pantryman, one trimmer and one helmsman.[36] The deserters were mainly seamen not firemen, despite the tough conditions under which the latter had been working in a heavily rolling ship. There were more opportunities for employment for seamen than firemen at this early stage of steam development, even though steam was used extensively on American rivers.

There had been a few other issues on this passage to New York. There had been hints previously given of problems with the compasses, which did not act 'as perfectly as could be desired. The variation was occasionally very considerable.'[37] It was on this passage that the ship encountered her first navigational problem when she touched a shoal off Nantucket. Hosken's report talks of a 'thick dark night' and the loss of the foremast.[38] A passenger letter gives more details and mentions a strong current that carried the ship into the shoals and that the ship was short of fuel. There was damage to the propeller.[39]

It is worth remembering the still experimental state of propeller-driven propulsion. When the Queen inspected the *Great Britain* on 23 April, she was shown three models of different propellers. There was a six-bladed propeller, which was the one that was finally fitted on the ship, a four-bladed reserve screw and another was a model with only three blades, the initial choice.[40]

The ship was eventually extricated from this unfortunate position on the shoals and limped into New York where she was dry docked, an expensive operation. On inspection, it was found that the propeller had been extensively damaged.[41] After repair the ship headed back to Liverpool, but the propeller continued to be problematic and by 6 November most of the propeller had now completely gone. On 8 November the engines were stopped with what was left of it vertical so as not to cause drag and the ship proceeded under sail.[42] The winds had freshened and there was a westerly gale speeding the ship on her way. The log describes the ship 'scudding and steering beautifully taking spray on larboard quarter and beam occasionally but is easier than any ship I ever knew'. On 11 November they took pride in passing two ships: 'This is wonderful with our little spread of canvas and more than I expected, well as I thought of her sailing qualities.'[43]

The naval architect, Corlett, in his groundbreaking book on the *Great Britain* sings the praises of that first propeller to cross the ocean.

It was a brilliant achievement that deserves remembrance. To fail because of the modern evil of metal fatigue and because of the mishandling of the ship was no disgrace; and those who smile at its short life would do well to reflect that very few real pioneers ever have done as well as Isambard Brunel and his colleagues did with this grand effort.[44]

During the winter alterations were made to pumps and valves to provide better air flow, which provided more steam and the ship was fitted with a new screw propeller.[45] So in the new season in 1846 the Great Western Steam Ship Company could truly demonstrate its ambitions with not just one steamship providing a regular service, but two. The local newspapers carried the advertised times of steam to New York. They provide a schedule of the sailings from Liverpool for both the *Great Western* and the *Great Britain*.

Gore's Liverpool General Advertiser, Thursday 12 February 1846 and 16 April 1846:

Great Western	1846		
Saturday	11 April	Thursday	7 May
Saturday	30 May	Thursday	25 June
Saturday	25 July	Thursday	20 August
Saturday	12 September	Thursday	8 October
Saturday	31 October	Thursday	26 November

Great Britain	1846		
Saturday	9 May	Saturday	6 June
Tuesday	7 July	Saturday	1 August
Wednesday	26 August	Tuesday	22 September
Tuesday	20 October	Tuesday	17 November

For the *Great Western* the advertised fares were 30 guineas plus the steward's fee of one guinea, since this ship carried just one class of passenger. The advertised fares for the *Great Britain* were 25 and 30 guineas, with the option of a single berthed room at 35 guineas plus the steward's one guinea fee.[46] Bookings for both ships were good and the future looked rosy. Passengers were attracted to the new ship but bookings for the *Great Western* remained strong, regularly attracting more passengers than the *Great Britain*, perhaps out of loyalty to a tried and tested vessel.[47]

The *Great Britain*'s second season commenced on 9 May but with a small complement of passengers, just 28. This increased on 8th June to 42 and the best numbers were on her passage from Liverpool in July when she carried 110 passengers. On this passage Hosken again had a navigation problem when, in thick fog, the ship touched a reef off Newfoundland. Despite this, the ship was showing real promise that she could make regular passages of 13 days out and 11 days back.[48] With both ships performing well, the future finally looked promising for the shareholders of the Great Western Steam Ship Company. That promise was not to be fulfilled.

4

'A USELESS SAUCEPAN': DISASTER ON AN IRISH SHORE

The Great Western Steam Ship Company now had two ships in operation and were hopeful of turning a profit after the extensive costs of the *Great Britain*. Captain James Hosken was at the height of his fame and was well known on both sides of the Atlantic, with wealthy and influential friends and contacts. The future looked promising for everyone concerned.

On the morning of 22 September 1846 in moderate weather the ship left Liverpool having on board 180 passengers, her largest passenger manifest yet, around 90 crew and a considerable quantity of freight, all promising future financial benefits for the cash-strapped company. Depending on the season and on the weather, from Liverpool, ships either ran south of Ireland or, as in this case, took the northern route. She passed south of the Isle of Man and would then plan to head north up the east coast of Ireland. Instead, in an inexplicable failure of navigation, she ran ashore on a rocky beach at Dundrum Bay in Ireland at 10 o'clock that night.

It was widely reported in almost every newspaper. The incident happened on the 22nd and it took a few days before the disastrous news reached England. Inevitably, Hosken endeavoured to show the positive aspects where he could. But nothing could disguise the scale of the calamity which had befallen the ship and the company.

Running Ashore of The Steamship *Great Britain*. (From our Correspondent.)

Liverpool, Thursday. – It is our duty to announce that the steam-ship *Great Britain*, which sailed from Liverpool for New York, under the command of Captain Hosken, RN., got ashore in Dundrum Bay, on the night of Tuesday last, about 9½ hours after she left this port.[1]

The first reports referred to the cause as the 'Commander mistaking the lights along the coast, and had not calculated he was so far on' and were quick to reassure that all passengers had landed safely and that the ship was still tight, not leaking, but stuck firmly on the shore. A short letter from Captain Watson, Secretary to the Underwriters at Liverpool, was posted in the underwriters' room giving brief official details:

Steam-ship *Great Britain*, Dundrum Bay, Sept. 23, 1846. Sir, – I arrived here in consequence of an express having been forwarded to me at Warrenpoint of the *Great Britain* being on shore off Tyrella, near the Watchhouse. I have only time to refer you to Captain Hosken's letter to Messrs. Gibbs, Bright, and Co., for more particulars. In haste to save the post, I am, &c Leonard Watson.[2]

Naturally the newspapers were eager to see Captain Hosken's letter, but Gibbs Bright rightly refused to make the contents known until the Great Western Steam Ship directors at Bristol had seen it. Gradually, more details emerged and were eagerly reported in newspapers across Britain and Ireland.

About half-past nine o'clock on Tuesday evening, the 22nd instant, the *Great Britain* went on shore in Dundrum Bay, between two rocks, very fortunately passing in between them on a soft sand. She is now lying there, the hull has sustained no damage, and she was not taking a drop of water at the time we last heard from her. The captain states that it will not be possible to get her off before the next spring tides. The passengers and crew were all landed safely. After she struck, the ladies, and a great portion of her passengers, were so satisfied of her safety that they went to bed, and remained on board until the following day.[3]

This all sounded reassuring and it was reported that after the passengers and crew had been safely landed, the majority were taken to Belfast, 'by means of jaunting cars, horses, carts, and other conveyances'. From Belfast some went onboard the *Sea King* and *Maiden City* steamers, which took them back to Liverpool and Fleetwood arriving on Friday, 25 September and the rest were expected the next day. Gibbs, Bright as the Liverpool agents were at full stretch organising passages on other steamers to New York. The ship had been carrying mail bags and these also were returned in the *Sea King* and were dispatched onwards to New York.[4] So relatives of passengers and crew were reassured, as were those who had sent mail. No mention was made of the cargo on board apart from passenger belongings. The anxiety now was among the insurers and the Liverpool Stock Exchange. They were reassured that the ship would be got off in the Spring. This was due to be her last sailing of the season, so little revenue would be lost.

The intelligence of the accident created a great sensation on 'Change, to-day, and those of the passengers who had come over on the Belfast steamers were eagerly sought after to learn the particulars from them. We are happy to have it in our power, by the publication of this edition to satisfy a multitude of inquiries and to allay the apprehensions of those who had friends and relatives on board. Lloyd's were informed that all crew and passengers were saved and the vessel was 'so far not injured'.[5]

Rumours abounded through lack of information and there was 'no truth in the report which prevailed in the town about an hour ago that Captain Hosken had arrived in Liverpool by the Newry steamer'.[6] Everyone was keen to hear from Hosken, but he remained with the ship valiantly trying to get her afloat.

A London newspaper gave more details and praised the actions of Captain Hosken.

> When the ship struck the passengers rushed on deck. The wind was in her favour, and she appears to have made most rapid progress. Through some inadvertence, not yet explained, either in the reckoning or in mistaking of a light, she got six miles out of her course; and at half-past eight o'clock in the evening, the passengers were alarmed by a sudden concussion, as if she had struck upon a rock. The shock was repeated, and they rushed upon deck. It was then ascertained that the vessel was aground. At this critical moment, Captain Hosken exhibited, it appears, the greatest self-possession. He apprised the passengers that the vessel was on shore, but that there was no danger. It was blowing hard at the time, but not so violently as to cause any apprehension. The signal of distress was made, the boats lowered, and assistance arrived from shore. Captain Hosken inspired all around him with his own confidence, and his orders were strictly obeyed.[7]

It highlighted that Dundrum Bay 'was the scene, a few years since, of one or two dreadful wrecks' and incorrectly reported that 'the vessel had received no damage whatever'.

> Captain Hosken remains with the vessel, and it is believed here that at the next spring tides she will easily be floated off by lightening her of the 800 tons of coal now on board. As yet, the only intelligence we have is derived from the passengers: and, as they are ignorant of nautical affairs, nothing positive can be ascertained of the cause of the disaster. The passengers represent the scene on board when the vessel struck as one of great confusion. The ladies rushed from their own cabin into that of the gentlemen, and some of them are described as 'huddled up terrified in a corner'.[8]

For the Great Western Steam Ship Company there was no disguising the total, very public and embarrassing disaster. It was a public relations catastrophe. In the days before newspaper photography the journalists had to do their best to conjure up the scene in words and inevitably the reports featured the dramatic and lurid. Artists depicted the scenes in drawings. Rolt, who wrote an early biography of Brunel, selected a particularly vivid account by one female passenger who was speaking to a reporter for the *Illustrated London News*.

> I cannot tell you of the anguish of that night! The sea broke over the ship, the wave struck like thunderclaps, the gravel grated below. There was the throwing overboard of coal, the cries of children, the groans of women, the signal guns, even the tears of men, and, amidst all, the voice of prayer, and this for long dark hours. Oh! What a fearful night![9]

The ship quickly became a tourist attraction. From Belfast Captain McLenaghan was offering a pleasure trip in his steamer on Sunday to Dundrum Bay 'when Passengers on board will have a fine opportunity of seeing the Iron Steam-Ship *Great Britain*, stranded in the Bay'.[10]

Acting on behalf of the underwriters, Captain Watson by now had seen matters for himself and was able to report:

> She now lies upright on the sand, and apparently has received no damage to the hull, with the exception of the rudder. In consequence of the wind shifting to the north-east last night the vessel did not alter her position, and must now remain until the next spring. Meantime Captain Hosken is arranging to sell all the coal on board, and otherwise to lighten the vessel as much as possible. I suppose the cargo will be landed, and reshipped for Liverpool. We will require considerable assistance in the way of steam tugs. All will depend on the state of the weather. I do not apprehend the vessel can receive any further injury. A steam-tug arrived this morning from Belfast, and is now employed securing out a bower anchor and chain the whole length to the E.S.E., in order to hold

the vessel, and prevent her getting further on shore. Every exertion is taken in order to preserve and save this fine vessel. Captain Hosken wrote fully to Messrs. Gibbs, Bright, and Co., yesterday. Captain Hosken speaks in praise of his crew since he left, and have no doubt they will continue to do their duty. I have nothing further to communicate on this truly unfortunate occurrence.[11]

He was then able to report the very welcome news that the ship did not appear to have sustained any significant damage. 'She was nearly dry all round this morning's ebb.' The critical matter was to secure her to avoid further damage and to unload the coal.[12] Newspapers were, meanwhile, still seeking first-hand reports from passengers.

We have been favoured with the following particulars by a gentleman of Montreal, who was on board the *Great Britain* when she went ashore: – He says that he was in bed and asleep at the time the vessel struck. He was awoke by the concussion, and, on jumping up, heard loud screams from the ladies' cabin. He ran on deck. The night was stormy, and the sea was breaking over the ship, which still continued drifting, and it drifted, he thinks, fuller a mile.

On the vessel striking, and during the remainder of the night, blue lights were burnt and guns fired, and the scene on board was most distressing. The great majority of the passengers were very ill: the Viennese children kept crying violently around Madame Weiss; and several passengers, who ought have shown firmness, betrayed lamentable weakness. Captain Hosken continued cool and composed, and several times referred to his charts. No one retired to rest – all were full of fears; but it was found next day that one passenger had never awoke until morning. At 4 o'clock in the morning boats came alongside, and several persons went ashore, and as the tide receded men waded alongside and carried passengers ashore on their backs; several, however, wished to remain on board, but Captain Hosken insisted on their leaving the vessel, as he could not be answerable for their lives. In reply to a formal application he advised them to repair at once to Liverpool, and apply to the agents. He could give them no money, guarantee nothing. Hundreds of men appeared to act as porters; but several American passengers refused to pay them, referring them to Captain Hosken, dispute and confusion were the consequence.

The coast-guard behaved admirably, protected property and passengers, and the wife of the watchhouse-keeper (the watchhouse was within 100 yards of the vessel) not only cooked for the ladies, but gave them all the provisions she had, and refused to accept one farthing in return. The passengers propose rewarding her. The country people around showed equal kindness, but the porters and carters were most exorbitant in their charges. The coaches charged only usual fares, and hotel keepers and publicans were equally moderate, but the housekeepers in Downpatrick charged a guinea for a bed.[13]

The reference to the Viennese was to Miss Weiss and her dancing troupe of 30 children, who understandably panicked. After leaving the ship they were helped by the local vicar who housed them in the school-house at Dundrum Bay. 'We hear she is in great distress, and is now in Liverpool, where she would perform last evening, for the benefit of herself and the troupe, at the Royal Adelphi Theatre.'[14]

Also on board the ship was a senior US politician and diplomat, William Rufus Devane King, who was deeply concerned about delay. A senator and previous vice-presidential candidate, he had been appointed as the American Ambassador to France in 1844. He was now in a hurry to return to America to contest his previously held Alabama Senate seat. He would later become the 13th vice-president when elected to serve with President Franklin Pierce in 1853, but he was already severely ill with tuberculosis. He was convalescing in Cuba and uniquely took his oath of office there, the only time this has happened outside Washington. He died just six weeks later.[15]

These were among the many angry and distressed passengers with whom the company and its agents had to deal. The company acted with commendable speed and it was announced that 'Messrs. Gibbs, Bright and Company as the agents of the Bristol Company which owns the *Great Britain*, agreed at once to refund the passage money. The passengers heard the news with much satisfaction, many of them being penniless.'[16]

Getting them on another steamer to New York, however, was not so simply fixed. MacIver, the Liverpool agents for the Halifax and Boston steamers, Cunard's line, were approached to see if the *Acadia* could be used. Around 60 passengers approached them and asked if it was possible as long as the passage money did not exceed £40. But MacIver declined to assist on the grounds that it might lead to a breach of their contract with the Admiralty, since any delay to the standard departure time meant a fine of £500 for every twelve hours' delay. This caused consternation as the *Caledonia*, another of Cunard's four steamers, due to sail for Boston on the next Sunday, was already fully booked. Many of the passengers were thus having to contemplate a forced wait until the spring. Those passengers destined for Havana could at least sail by the first steamer from Southampton.[17]

Claxton had rushed to see the ship for himself and to get a first-hand report from Hosken of the circumstances. She was in one piece but had settled on some rocks, which had knocked holes into her. Now most of the passengers were back and there had been negative press reports of missing luggage. From Dundrum, Claxton wrote a short letter to *The Times* on 28 September, keen to correct those reports and defend the local people.

Sir

A report having gone the round of the papers injurious to the character of the country people near Dundrum Bay, I feel it my duty, and it is with much pleasure that I inform you, that nothing has occurred since the *Great Britain* came onshore that can at all justify such a report. The natural degree of eagerness to obtain the luggage between boatman and men with carts, for the payment for the same, and a concomitant confusion, occurred, as was likely to occur; but it is more probable, if anything has been missed by any of the passengers, it has been owing to that confusion than from premeditated dishonesty.

To the poor as well as to the rich in the neighbourhood the grateful thanks of the directors are due, for nothing can exceed the desire to alleviate the misfortunes of all concerned that has been manifested by everyone in this locality. Your giving currency to this will oblige.[18]

He then followed this with a separate letter, this time in defence of Captain Hosken and his navigational skills. Claxton had examined the log and the chart. The chart in use was published by John and Alexander Walker, agents of the Admiralty in Liverpool. It had been purchased in June of the previous year as the 'latest and most correct chart', but did not have any mention of a light on St John's Point. It was this new light which was blamed by Hosken for the confusion.

Claxton praised Hosken's actions since the stranding, expressing his 'admiration of his manly bearing under circumstances that might well have crushed a weaker mind and of the discipline and good conduct of the officers and men under his command, who are working gallantly, I may say, under circumstances of great trial in difficulty if not of danger'. He reasserted that the compasses were perfectly correct and praised the construction of the ship, which was able to withstand the strong waves and high seas that would have broken up 'the strongest wooden ship that ever was built'. He finished with the cheerfully optimistic statement that 'if we are favoured with tolerable weather I see at present no reason to doubt her being afloat by the end of the week'.[19]

The weather did not co-operate. The following Monday in the higher spring tides Claxton tried to get the ship off but a gale of wind prevented the operation and the decision was taken to drive the ship higher up the beach where she would be slightly less exposed. 'Sails were therefore set and she was driven forward a considerable distance.' This must have been some sight, the great iron ship under full sail heading further into the bay. The shipbuilder Patterson was now sent by the directors to Dundrum together with Mr James Bremner, who was a salvage expert with specialist experience in floating stranded ships. They set about creating timber breakwaters to protect the ship on the exposed shore, but these were unable to withstand the gale force winds.[20]

The cause of the accident continued to be the subject of public debate. Hosken had blamed inaccurate charts and his explanation was supported by the company. Others suspected it was due to compass error. Since 1835 there had been awareness of this problem when Captain Johnson showed the Admiralty the impact of an iron ship becoming a large magnet and distorting the needle. The Astronomer Royal Professor Airy, a distinguished mathematician, came up with a complex solution for adjusting the compasses on each iron ship. The adjustment was done by agents in each port.[21]

One passenger on board the ship at the time of the accident witnessed some debate about the compasses:

There was a light visible, and Captain Hosken said it was about a mile and a half distant, and, in reply to question, said that they were near Ardglass. An officer of the ship was examining the compass, and the Rev. Mr Tucker, of Bermuda, seeing that it did not work, asked him the reason. He replied, after some time, to the effect that there was something the matter with it. The same officer then examined the compass in the fore part of the ship, and an opinion prevailed that the compass was the cause of the disaster. Hearing this, Captain Hosken solemnly declared that there was nothing wrong with the compass.[22]

Correspondents wrote to the newspapers in defence of Captain Hosken. 'The character of Lieutenant Hosken has long stood too high, both in the Navy and in the Merchant Steam-service, for any one conversant with either to suspect that the unfortunate loss of the *Great Britain* was in any way caused by want of skill or care on the part of her commander.'[23] The writer blamed a faulty unofficial chart. Charts could be produced by anyone and Admiralty charts, from the Hydrographic Office, had apparently not been in use onboard the ship.

'Captain Hosken is acquitted of all blame by Captain Claxton. secretary to the company. It is stated that one of the coast lights was not marked in Captain Hosken's chart, although it is one of 1846.'[24] While the company directors loyally blamed navigation charts, other commentators put the blame on Captain Hosken. The *Mechanics Magazine* was firmly of the view that 'the case was on his own showing and beyond all possibility of doubt the most egregious blundering.'[25]

Hosken, himself, was dubious about the accuracy of the compass. Just before the fateful passage, on 8 September 1846, he had entered into private correspondence with the Astronomer Royal, Professor Airy. Hosken described frequent problems and asked about local influences which might cause compass errors during a journey. Writing to Hosken after the wreck, in November, Airy acknowledged the possibility of misleading variations in adjusted compasses. This correspondence was never made public.[26]

Hosken's report dated 13 October 1846 showed how he had been working unsuccessfully and exhaustively to get the ship off the beach and stop the leak by the stern since the incident. He explained the circumstances of the stranding and his decision to take the northern passage around Ireland rather than head south past Holyhead. He described the situation on board and how he calmed the passengers, having ascertained that there was no danger of loss of life, and had them remain on board until daylight. Then lifeboats were lowered and passengers were landed with some of their luggage. In his report he mentions 'a great rush among the poorer people for jobs for which they took good care to be well paid'. The stranding of the ship with its many passengers was something of a brief bonus at the time to the Irish poor who were suffering considerably through the severe famine. Local officials and local people were of great help and he highlighted the coastguard led by Captain Fisher and the officer in charge, Lieutenant Morris. He singled out Lady

Matilda Montgomery. who gave shelter in her own home to several passengers and many more in cottages locally. The local magistrate, Captain Despard, also assisted in securing items and helping with any local disputes, as did the Constabulary Force. Captain Hosken took the blame firmly on his shoulders. 'I take the blame, if there be any, to myself, for it is I alone who had anything to do with the navigation of the ship, or any other I have ever commanded.' He was mortified by the whole incident and presented his hitherto blameless career for the consideration of the company.[27]

The company circulated Hosken's report to the proprietors and sent special thanks to all of the various people named in his report. The directors firmly blamed the omission of the notice of St John's light in the chart and they planned to appeal to 'all insurance companies, merchants, shipowners, and masters of ships to aid them in urging upon the legislator an enactment with heavy penalties against all Admiralty agents' who sold out-of-date charts.[28]

The other aspect of public relations was to reassure all and sundry of the safety of the iron ship. 'Strong hopes are entertained that the ship may be rescued in the course of a few days. It is mentioned as a proof of the admirable qualities of the ship, that the doors throughout the ship opened and shut within the frames with perfect accuracy.'[29] It was better to blame an error in a chart than to allow a suggestion that the iron ship had affected the compass. One of Brunel's biographers believed it was Brunel's personal view that it was a compass error.[30]

The Great Western Steam Ship Company had been experiencing financial difficulties for some time, largely due to the ever-increasing cost of building the *Great Britain*. They had pinned their hopes on securing subsidy through a government mail contract to no avail. Although like many ships at the time they carried private mail, government contracts for official mail were valuable subsidies. The Cunard line had retained the Atlantic contract despite several appeals from the Great Western Steam Ship Company.[31]

Now the concerned shareholders raised the question of insurance. In the past at other meetings they had been content to self-insure, but now when good insurance could save the day there was considerable anxiety to discover exactly what cover was in place. At a private meeting of some of the proprietors of the Great Western Steam Ship Company, the direct question was asked: was the ship fully insured or not? The chairman confirmed it was not, apart from insurance on a sum of £18,000 to 'secure Mr Miles's mortgage to that amount on the vessel, as there was some difficulty or difference effecting the insurance'. The insurance difficulty related to a lack of a classification from *Lloyd's Register*. Lloyd's surveyors were very cautious about experimental ships. Miles, a long-term shareholder and director, had clearly provided a loan of £18,000 in the form of a mortgage to finance the ship. It was pointed out by one of the shareholders that they should have been informed of the lack of insurance so that they could, as individuals, take out their own insurance to the extent of their share or shares. As this had not happened, some of those present felt that they 'may hold the chairman or directors responsible for their loss' and discussed obtaining legal opinion.[32]

One month later an aggrieved shareholder wrote to the *Bristol Times and Mirror* suggesting that there was still some considerable confusion over the status of the insurance and referring to the charts. 'It is so replete with curious information as to how the dotting down the red ink marks in the old charts might be effected.' This suggests that in dealing with underwriters, even for the relatively small sum of insurance, the cause of the accident was being carefully handled. One week later in another letter the same shareholder asked, 'What became of the *Great Britain*'s stores? What became of the passage money? Is Captain Hosken still in receipt of his salary? And what have the Directors done?'[33] Information was limited and feelings ran high.

The 1840s saw Brunel at his busiest with work on bridges, railways, docks and parliamentary committees and he was also dealing with the problems of his atmospheric railway in Devon.[34] But, at last, on 8 December he could travel to see his ship as she lay stranded in Dundrum Bay. Expecting the worst he found the ship 'almost as sound as the day she was launched, and ten times stronger in sound and character'. Hosken was still there and Brunel instructed him to work on a new way of protecting the ship, personally guaranteeing that he would cover the expense of the initial work in the event of the directors not agreeing. He wrote a lengthy letter to Claxton on his return and his opening sentence was that he had 'very mixed feelings of satisfaction and pain, almost amounting to anger, with whom I don't know'. And he was furious that 'the finest ship in the world, in excellent condition, such that £4,000 or £5,000 would repair all the damage done, has been left, and is lying, like a useless saucepan kicking about on the most exposed shore that you can imagine with no more effort or skill applied to protect the property than the said saucepan received on the beach at Brighton'.[35]

He was then concerned about the ownership situation: as to whether it was the company, which could then act quickly, or whether it had to go through the underwriters. His letter is full of his passion for his creation, his determination to save it – and a neat variation on the bolting of stable doors.

As to the state of the ship, she is as straight and as sound as she ever was, as a whole. I told you that Hosken's drawing is proof, to my eye, that the ship was not broken: the first glimpse of her satisfied me that all the part above her five or 6 feet water line is as true as ever. It is beautiful to look at, and really how she can be talked of in the way she has been, even by you, I cannot understand. It is positively cruel; it would be like taking away the character of a young woman without any grounds whatever.

The ship is perfect, except that at one part the bottom is much bruised and knocked holes in several places... For really when I saw a vessel still in perfect condition left to the tender mercies of an awfully exposed shore for weeks, while a parcel of quacks are amusing you with schemes for getting off, she in the meantime being left to go to pieces, I could hardly help feeling as if her own parents and guardians meant her to die there... What are we doing? What are we wasting precious time about? The steed is being quietly stolen while we are discussing the relative merits of a Bramah or a Chubb lock put on at some future time! It is really shocking.[36]

The strain, especially on the stern of the ship, had be reduced and a stream of water, which was running underneath her, had to be diverted before it washed away the sand. To protect the ship from the power of the sea on the exposed shore, he set Captain Hosken to work on the new arrangements for protecting her with a temporary breakwater.[37] Brunel then wrote a rather more considered report to the directors. In this he pointed out the significant difference between a wooden and an iron vessel and urged them to put all thoughts of their experience of wooden shipwrecks from their mind, as this was a wholly different case. In his view, rather than trying to break up the ship, a massive task for an iron-built vessel in a remote part, the ship should be rescued and brought into port where he felt that even unrepaired she would be worth between £40,000 and £60,000.

But in the meantime the urgent matter was that *'the ship must be protected.'* And he recommended the system of banks of fagots made of strong and long sticks packed closely and in great thickness up against the sides of the ship, bound tightly to the vessel with chains. It was a system practised in Holland for protecting the repair and protection of banks against the sea. He urged them, 'You have a valuable piece of property lying on the most exposed shore.' If neglected, he pointed out, it would be worth nothing and he appealed directly to their commercial experience.

Can you as men of business under such circumstances waste your time at this moment in discussing what you are doing three months hence, what plan you will then adopt to take your property to market, but will you not rather first and immediately adopt decisive steps for preserving that property and then consider what you had best do with it?[38]

Brunel finished by praising Mr Bremner, the salvage expert, and saying, 'I firmly believe that if any man could take her off (and it would be prudent to let him do so) Mr Bremner's great experience and sound practical knowledge and good sense in devising any plan, and his energy and skill carrying it out would ensure every chance of success.' His final strong recommendation was that Claxton should be put in charge of superintending the protection of the ship during the winter months together with Mr Bremner.[39] This would relieve Captain Hosken.

Claxton duly headed across, and as secretary of the company was probably more than happy to be out of the way of furious passengers and anxious shareholders and doing something more practical. There followed a constant stream of letters between him and Brunel. Creating the breakwater was not a simple task. Vast numbers of wooden sticks had to be bound together and secured around the ship before the strong seas swept them away. Such was their relationship that Brunel could be extremely direct. On 29 December 1846 he wrote to Claxton 'You have failed, I think, in sinking and keeping down the fagots from that which causes nine-tenths of all failures in this world, from not doing quite enough.' He urged him to stick to the same plan but just do considerably more of it in creating the fagot bundles weighed down by sandbags. He finished by urging if a 'six bundle fagot won't reach out of the water, try a 20 bundle one; if hundred weights won't keep it down, try tons'. It was no simple task as these bundles of fagots averaged 11 feet long and 5 feet wide and were cumbersome to handle. One of Guppy's 30-foot iron lifeboats was loaded with stones and sunk by the side of the ship to form a base.[40] The ship's cannons were also used as anchors for the breakwaters.[41] Frequently heavy gales might wash away the work, only for the sailors and other workmen to replace them at the next low tide.[42] By 27 February 1847 the ship was protected and Brunel had warm praise for Claxton when writing to the directors: 'I had relied confidently on success when my friend Captain Claxton undertook the work, and the result has fully confirmed my expectations.'[43]

There was no shortage of offers of help from other quarters or ideas to help raise the ship. 'The Great Western Steam Ship

Company have filed at their offices no less than 460 plans, which they received from various quarters for floating off the *Great Britain*.'[44] These came from engineers, enthusiasts, the fanciful of mind and even one from a young boy anxious to help. Brunel's office received such a volume of helpful suggestions that he had a standard letter printed as a reply to decline assistance.[45]

The stranding of the *Great Britain* sealed the fate of the *Great Western* steamship. Due to the rising cost of the building of the *Great Britain*, it had been on the market for a while but now the company urgently needed to raise funds to salvage their new flagship. The Royal Mail Steam Packet Company had the mail contract for the West Indies and needed new ships. Their senior man was Captain Chappell, who had previously been the master of the celebrated screw steam ship *Archimedes*, so he knew the company well. In 1847 the *Great Western* was sold for £25,000 and taken from Bristol to her new base at Southampton. The Royal Mail planned to spend a further £10,000 on refitting her before putting her into service.

The world looked on as the salvage party worked to rescue the great iron ship. Every heavy item was removed including the engines. The *Illustrated London News* carried a constant series of drawings with news of progress. Having protected the ship for the winter months, the very big challenge now was to lift the ship to make a temporary repair and get her back. Claxton together with Bremner and Bremner's son, Alexander, set to work to follow Brunel's bidding and sent daily reports back to Brunel who had every faith in the team. The ship was levered up and, as it was lifted, stones and wedges were rammed underneath to maintain her position. At last the ship was raised enough to allow access to the bottom and make it watertight.[46] It was a highly challenging, difficult and delicate task, but at last, on 27 August, 11 months after her stranding, Claxton could write in triumph to Brunel.

Huzzah! Huzzah! You know what that means. I made my mind to stop at the edge of low water, and then examine and

secure all that might discover itself. The tide rose to 15 feet
8 inches. She rose therefore easily over the rock, but was clear
of it by only 5 inches, which shows how near a squeak we
had – it was a most anxious affair, but it is over... I have no
doubt that tomorrow we shall see her free.[47]

The ship was finally released from her enforced imprisonment
on the beach, while ironically promoting the durability and
strength of iron as a shipbuilding material. Such was the
importance of this vessel that the Admiralty sent two ships,
Birkenhead and *Scourge*, to assist, plus men from the dockyards
at Portsmouth and Plymouth. With considerable effort from
the men who were drafted on board to pump out the water
the leaking ship arrived in Belfast, where she was temporarily
grounded. The next day with the pumps working the whole
time the ship was taken across to Liverpool. It was a spectacular
marine salvage event.[48]

Meanwhile, the company, having saved the *Great Britain*,
had to realise that this was the end not only of the work but for
itself as well. The ship was basically repaired and then put up
for sale. At the next annual general meeting held in their offices
in Princes Street, Bristol, with Henry Bush in the chair, Claxton,
as secretary, read the report of the directors. The *Great Western*
had been sold for £25,000. The cost of the protection and
removal of the *Great Britain* after deducting the proceeds of
materials sold, amounted to the sum £12,670 14s 1d., and it
was estimated to cost £21,694 to restore her to her pre-accident
condition. An agreement had been reached with the Liverpool
underwriters, and a similar agreement was anticipated with the
underwriters in Glasgow.[49]

Offers to purchase the ship had been made but could not be
acted upon until the insurance situation was settled. It had been
rumoured that the company would repair the ship itself, but this
was denied by the directors, who firmly stated their intentions
to wind up the company as soon as possible. Once the insurance
had paid out, they expected to be holding £7,000 or £8,000.
This would be returned to the owners of the new shares at

£5 per share. The old shareholders, who had benefited during the time when the company was earning well, had nothing. As other sums were realised from the sale of assets such as the *Great Britain* these would also be paid out. The dockyard was let to a Mr Hennett on a five-year lease, breakable at the end of three years and so they planned to give six months notice to the tenant. There was still the plate of both company ships, which might earn them £1,400 to £1,500, and also some premises in New York. These, with the cash in hand, the claims with the underwriters, the *Great Britain*, the works, and certain premises in New York and Liverpool, constituted the whole of the assets of the company.[50]

Mr J. W. Miles moved the thanks of the proprietors to Mr Brunel for his highly important services, and to Captain Claxton for his energy and ability in removing the *Great Britain* from Dundrum Bay. With some satisfaction those in the the room were happy to record their thanks for the rescue of the ship and the efforts required to do so. Mr Abbot, who admitted he had once been a vociferous critic of the directors, now rose to praise the actions they had taken to free the company from its liabilities and to compensate the proprietors.[51] The shareholders/proprietors still had considerable liabilities as the company was still not incorporated. The winding up of the company and the sale of all the assets released all the owners from those liabilities, which explains the sense of relief at the meeting.[52]

At the annual general meeting in 1849 the company still had the *Great Britain* on its hands, and it was costing money as it remained in Liverpool. They announced a potential plan to operate the now engineless ship as a sailing ship between Liverpool, New York and New Orleans to carry emigrants and cotton, but no action was taken.[53] After a few failed attempts she was finally sold for the bargain price of £18,000 in December 1850, which incidentally was equal to the mortgage held by Miles. It was initially announced that the ship had been purchased by William Patterson, the shipbuilder, but he was acting for Gibbs Bright of Liverpool. The Gibbs, Bright company had been the ship agents for the *Great Western* and *Great Britain* in Liverpool.

Their interest in the ship was for new opportunities opening in trade with Australia.

The final directors' report was in March 1852 confirming the sale of the *Great Britain* and the reassignment of the company's lease on the Bristol dock and the works to Patterson.[54] Following the end of the Great Western Steam Ship Company, the key figures who had been so much part of the creation and careers of the two great ships went their separate ways. The man whose career was severely damaged was Captain Hosken and in his memoirs he admitted that 'Dundrum Bay, was the end of a singularly successful career for many years.' It was an extremely difficult time for him personally and his fall from such a prominent and well-regarded status was swift and humbling. 'It also proved to me who were true and who were fair weather friends. The repeated kindness and consideration shown to me on many hands were a great alleviation under the circumstances.'[55]

In 1847 Hosken made a delivery run of a steamship for a friend to New York and after a week or two there he returned on the Cunard steamship, *Caledonia*, from Boston, 'the owners having very liberally offered me a free passage to Liverpool'. With the help of other friends he obtained the position of the post of harbourmaster, postmaster and chief magistrate of Labuan in the Far East, where he spent two years, in self-exile, well away from comment. This was a terrible fall for a man who once could have secured a position at the helm of any of the world's greatest ships. He was still on the Navy List and, perhaps with the help of sympathetic friends, in 1851 he became commander of HMS *Banshee*, a naval packet boat in the Mediterranean. When the Crimean War broke out in March 1854 he was appointed to HMS *Belle Isle*, which became a hospital ship for the Baltic fleet and, after a brief time in the Crimea in 1856, Hosken finished his time with the navy.[56] The Dundrum Bay incident remained an albatross around his neck for the rest of his career, but he was able to regain some of his status when, in retirement, his gradual promotion in the navy meant that he at last reached the status of Admiral.[57] In his retirement he wrote his memoirs, which were published after his death by his second wife. Like many memoirs they were his attempt to justify his career.

Thomas Guppy's career also faltered. As a director of the company in 1841 he stepped in to become the Engineering Director and take on the task of finishing the *Great Britain* and, with the company unwilling to pursue further projects for themselves or other parties, Guppy's presence as part of the company was reduced.[58] He did build a small vessel in the company's yard on his own account. The *Richard Cobden* was launched in 1844, an iron barque of 450 tons, 156 feet long and 27 feet wide, built with 'apparently excellent constructional design materials and workmanship'.[59] Guppy then disappears in connection with the company's affairs and reappears as a trustee of the Coombe Avon Company of Copper Miners in Glamorgan, which was subsequently bankrupted in November 1848.[60] The following year he resigned as a director of the Great Western Railway.[61] There may have been financial concerns as ten years later there is a sale of a life insurance policy on him for £8,000.[62] He moved to Naples for health reasons and there established a very successful engineering firm described in later years as affording the 'means of subsistence to six hundred families'. He retained an interest in marine engineering, designing and constructing a torpedo boat for the Italian government and remained there until his death in 1882.[63]

Guppy's reputation in Bristol remained high and in later years he returned to be a part of the company to finish the Clifton Suspension Bridge. He was remembered fondly:

Of late years we have lost sight and almost recollection of the once widely known and talented citizen, but the announcement of his death recalls attention to him; there are, we expect a few yet amongst us who, like ourselves, retain a recollection of his intelligent face, his readiness of purpose and his active and energetic demeanour.[64]

The rescue of the ship from Dundrum Bay enhanced the reputation of the civil engineer and salvage expert, James

Bremner. In 1901 a fund was established by the Town Council in his home town of Wick, Scotland, to build a statue in his memory, 'the raiser of the *Great Britain* steamship'.[65] As for Patterson, the shipbuilder, Christopher Claxton, the managing director of the Great Western Steam Ship Company, and their consulting engineer Brunel, their relationship with the *Great Britain* was not quite over.

5

'THIS MAJESTIC VESSEL, QUEEN OF THE OCEAN': NEW OWNERS

Gibbs, Bright, now the owners of the *Great Britain*, were not just ship brokers and agents but also owners of the Eagle Line of sailing ships trading with Australia. The Australia run was hard on sailing ships. The ships ran with the trade winds down the Atlantic almost to South America and then turned east towards to the Cape of Good Hope and here in the southerly regions they could encounter high seas, strong headwinds and ice. The *Great Britain* was a tough, well-built ship with good sailing qualities. She was also very large and could potentially accommodate a great number of passengers.

The United States was by far the most popular destination for emigrants. Between 1815 and 1859 nearly three million people travelled across the Atlantic from Britain and Ireland, with half that number going to Canada. When the *Great Britain* was purchased by Gibbs, Bright in December 1850 the emigration numbers for that year were United States 230,885, Canada 32,961, Australia 16,037.[1] Much of the emigration to Australia was financed by government charter, some migrants paid their own fares themselves or were supported by charitable organisations.

Emigration to the Americas was much more likely to be self-financed.[2] Liverpool's Australian trade had a complex pattern and in mid century much of this traffic was concerned with emigrants rather than goods. Direct inward traffic from Australia to Liverpool was rare; most sailing ships came back via ports in India or headed across the Pacific to South America where it was easier to obtain return cargoes.[3] The *Great Britain* being mainly passenger and some cargo could change that pattern.

The discovery of gold in Australia in Victoria in 1850 had a dramatic impact and there was a rush of emigrants keen to make their fortune in the gold fields. Ships of every description were needed to cope with demand. Gibbs, Bright already had good Australia connections and the first ship to arrive in Liverpool with gold from Australia was one of their Eagle Line packets, *Albatross*, landing £50,000 worth of gold dust. As Corlett says, 'Casting around for tonnage, Gibbs Bright who had been the *Great Britain*'s agents on the Atlantic run thought of the ship. Here was a magnificent vessel, the largest in the world, capable of taking considerable numbers of passengers – not to mention cargo – each way and going cheap.'[4]

Conversion of the ship to carry emigrants meant considerable rearrangement of her accommodation from first and second class only to add a space capable of carrying large numbers of people in third class or steerage. The engines were removed to be replaced by auxiliary engines. She was now to be a sailing ship with engines, as the plan was to use sails for much of the time to conserve coal and the new engines by John Penn and Company provided power only when required.[5] Steam was only economic on certain routes for a large part of the 19th century, and tended to connect a relatively small number of key locations on fixed timetables, in comparison with the old sailing ship patterns, which were much more universal in their coverage.[6]

William Patterson, the shipbuilder, who had been so involved with the ship since her inception, was put in charge of the conversion and Captain Claxton's services were also retained by Gibbs, Bright. With a long passage to Australia significantly more cargo space was needed for emigrant luggage. The total number of passengers was now 730, of which 50 were first class. There were ladies' boudoirs for privacy and an innovation was the hurricane deck, a pleasant space in tropical climes.

The grand saloon was 75 feet long and the *Illustrated London News* approved of the tasteful decorations. This saloon was also very popular in the colder climes, since it was one of the warmest spaces on the ship.[7]

By March 1852 an important event was reported, the adjustment of the ship's compasses. That this action merited special attention is notable. In light of the controversy over the stranding of the ship, this event was well publicised, but there was also a wider dispute between the Astronomer Royal, Professor Airy, who promoted his system of correcting compass deviation in iron ships and a ship's captain, philosopher and evangelical clergyman called William Scoresby. Scoresby was the chaplain to Liverpool's Mariners Church, 'a decommissioned man-of-war floating in Liverpool Harbour'.[8] Scoresby was in dispute with Airy and had sought to undermine his system. Since the stranding of the *Great Britain*, Scoresby claimed that the 'ship's magnetism might change drastically in response to events in the open sea thus undermining the efficacy of Airy's technique'.[9]

Airy's mathematically complex compass calculations could only be carried out by the instrument makers in port, especially in Liverpool.[10] His technique involved no knowledge or skill on the captain's part, but that, some commentators argued, made the captain 'less able to make an informed decision about the true position of the ship'.[11] Airy had strong views on 'the stupidity of captains' and sought to distance them as far as possible from anything to do with the compass, which cannot have helped in the dispute between their methods.[12] In Captain Scoresby's view, the magnets used in Airy's technique rendered the compasses 'worse than useless'.[13]

Against this background, and with the ship's history still fresh in many minds, it was vital to reassure the public about safety and so the local newspapers duly announced, 'This splendid craft left the quay at which she had been for some weeks laying, for the purpose of having her compasses adjusted by observation. The operation is a most interesting and important one.' Swinging the compass involved swinging the ship by slowly turning it in a circle stopping at each point of the compass.

Watching the proceedings onboard were Robert Bright, Samuel Bright and Claxton, 'to whose great exertions at Dundrum Bay may be attributed the preservation of the ship'. The adjustment was carried out by John Gray, Liverpool instrument maker. All were 'were delighted to find that, despite the amount of local attraction in this leviathan ship, her compasses were rendered as perfect as those of any craft afloat'.[14]

Later in 1855 on a voyage financed by the Liverpool Underwriters Association, Scoresby travelled on the first voyage of the *Royal Charter*. This ship, owned by Gibbs, Bright, was to be a sister ship to the *Great Britain*. Scoresby recorded changes in the magnetic character of the *Royal Charter* reporting the changes as he went. These drastically changed over the course of the journey and it was a triumph for Scoresby's views, but he died before he could publish anything. It would be many years before it was established that ships might change their magnetic character soon after they were built, only to stabilise shortly thereafter and be 'largely free of the problems of retentive magnetism that Scoresby had attributed to them'.[15]

In 1852, after this public demonstration of her compasses the *Great Britain* then proceeded to her engine trials. The fame of the ship meant that London newspapers were still very interested in reporting on her. The ship headed for Holyhead and the engines were reported as having 'worked beautifully'. Among the many people on board were Mr Penn, the maker of the engines, Mr Harman, the engineer on the original engines now based in London, together with Pettit Smith, Patterson, Claxton, Captain Kilsado of the Spanish Navy, Captain Bibby, Samuel Bright and Tyndall Bright.[16]

Claxton's report to the new owners after the trial, yet again took pains to highlight the compass performance.

In compliance with your request that I should note and report to you on the performance of the *Great Britain* and her engines, on her trial trip just ended, I have to acquaint you that I am enabled, with more than ordinary satisfaction, to state that everything has gone most satisfactorily, and judging the future by the past, I have no doubt that all the mechanical and naval anticipations will be fully realised...I have witnessed many experimental trips, but none in which the engines have given so little trouble – steam always being exceedingly plentiful, notwithstanding the inexperience of the stokers. We had but little opportunity of trying the ship under canvas, as the wind hauled us before we got clear of the sands; while, however, it was fresh and on the beams, the ship proved herself to be remarkably stiff. I am happy to say the compasses were as correct as they could have been in any steamer or sailing vessel of wood, which is highly creditable to Mr Gray, by whom they were adjusted. With respect to the screw, there is no second opinion on board as to the vibration being less than in any other steamer, and I have no kind of doubt that, had the ship been loaded, she will be in the screw from 2 to 3 feet deeper in the water it would scarcely be felt at all. The accommodation I will leave for those who are to be her future passengers to speak of; for my part, I am struck with the excellent and comfortable way in which everything on board has been arranged.[17]

The newspaper added the overly optimistic and erroneous forecast that 'the *Great Britain* is ... capable of carrying 3,000 tons cargo, in addition to 1,300 tons coals, which will do her out and home'.[18]

Claxton's profile and reputation was high and he was described as the 'eminent engineer'.[19] One year later his words were used in advertisements to endorse John Gray's floating compass.

IMPORTANT TO IRON STEAMERS AND other craft. Captain CLAXTON. R.N., has addressed the following letter Mr John Gray. Nautical Instrument Maker, Liverpool: –

The *Great Britain* Steam-ship, July 5, 1853.

Sir, I have watched with great Interest the Compasses of the *Great Britain*, during the successful experimental trip this day composed. Fortunately there was strong breeze, and when the ship was clear of Holyhead a good deal of quick motion, with her speed over twelve knots. I am decidedly of the opinion that your 'Floating Compass' is the very best I have ever for steam ships, where vibration is apt to affect these important articles. The Floating Compass had no vibration whatever, and your other Compass was by no means so much affected as the common ones.

I have great pleasure in expressing this opinion and I trust, if you think it of service, you will avail yourself of it, which the least I can for a gentleman who has done so much for navigation.

Your obedient servant,
CHRISTOPHER CLAXTON, Com., R-N-
Mrs Janet Taylor, 104, Minories, is appointed Sole Agent for London.[20]

With such public affirmations of safety and support from the major figures involved in her construction and refit, the *Great Britain* was now ready for her inaugural voyage under new ownership. This was not to Australia, but to New York. It has been assumed that the *Great Britain* was purchased for the Australia run but, with higher emigrant numbers on the Atlantic, it is just possible that Gibbs, Bright were tempted to use her there. Indeed this had been an idea floated within the last years of her ownership with the Great Western Steam Ship Company. A run to New York could both test the engines and rig and test the market. Barnard Mathews, who had served as second officer, then master of the *Great Western*, was appointed captain, following yet again in the wake of Hosken. The advertised fees were 20 guineas for the saloon and a few midship berths at 15 guineas. These included the steward's fee, the attendance of an experienced surgeon, bedding and all provisions, except wines and liqueurs 'which will be supplied at very moderate prices'.[21]

With 180 passengers on board, the *Great Britain* sailed again for New York doing well despite strong gales and arriving in 13½ days.[22] The passengers were mainly British and Irish. There were 100 English passengers, 31 Irish passengers, 12 passengers from Scotland and a mix of occupations: merchants, farmers, doctors, priest, artist and civil engineer, so they were not all from the labouring classes. The artist was George Inness, an American landscape painter, age 27 travelling with his 19-year-old wife Elizabeth. He was already a recognised artist, his talent had been seen at a young age. By the mid-1840s, he was studying at the National Academy of Design in New York and in 1851, Inness visited Europe for the first time. He spent time in Rome and Paris before returning to North America aboard the *Great Britain*.[23]

There was the usual laudatory letter signed by all the passengers praising the ship, its performance, 'the spirited proprietors, who have so efficiently restored to Atlantic navigation one of its finest ornaments', the new and improved machinery and in praise of the 'very satisfactory conduct of your officers and crew' and to the captain's 'unceasing vigilance and ability in command'.[24] Captain Mathews must have been a happy man.

The test run to New York cannot have brought sufficient return so in July she was widely advertised for Australia. Promising a passage 'in the shortest possible time' and referring to her New York run as a trial, conducted with 'remarkable rapidity', she was described as combining 'all the advantages of a clipper ship with those of a powerful steamer'. The fares to Melbourne for the after saloon were 70 guineas and upwards, for the fore saloon, 40 guineas and upwards, for the second cabin 30 guineas and upwards. No fares were mentioned for steerage.[25] The response was good, 630 passengers booked passage with the *Great Britain* and she carried a crew of about 150 and a very considerable amount of mail. It was the largest complement of passengers she had ever carried. It was anticipated she would reach the Cape in 25 days and then Melbourne in 56 days, halving the time of the average sailing vessel. In case of any attack the ship had six deck guns and was well supplied with arms and ammunition.[26]

The master, Barnard Mathews, was a Cornishman from Penzance who had spent all his life at sea on merchant ships before he joined the Great Western Steam Ship Company in 1837. He had been the master of a small steamship trading across the Bristol Channel, owned by Guppy. Mathews was Hosken's number two on the *Great Western* and eventually succeeded to the command of that ship.[27] When the *Great Western* was sold to the Royal Mail Steam Packet Company in 1847, Mathews became master of the *City of Glasgow*.[28] This was a new screw-propelled steamship ship purchased in 1850 by William Inman of the newly established Liverpool and Philadelphia Line and who would later use it as an emigrant ship.[29] Mathews left the Inman line to become master of the *Great Britain* after its transformation by Gibbs, Bright.

The first voyage of the *Great Britain* to Australia began on 21 August 1852. This was a very different voyage for the ship, instead of two weeks this would take months and pass through extremes of temperature. On board, the first- and second-class passengers would be joined by hundreds of passengers travelling in steerage. Passengers and crew were bound together in their floating world dependent on the qualities of their ship, crew and master in uncertain weather. Added to this, for many, was the uncertainty of their future in an unknown land.

Like most things it all began with good intentions. Mathews' concern for his passengers was noted. 'Captain Mathews was very anxious about the comfort of his passengers creating a sort of platform on the bulwark for the gentleman to lounge or play games, leaving the deck free for the ladies and children. This platform is a great convenience, and kept the bulk of the passengers clear of the deck. It was also better for the crew, but gave them more room on deck to navigate the ship.'[30]

On the first Sunday, the church service, an important event in the week for pious Victorians, was held in the main saloon and Mathews led the prayers but rushed off saying he had no time to give the sermon, promising one that evening, a promise he failed to fulfil.[31] This is completely consistent with his previous Atlantic crossings and, in fairness, rather depended on the strength of the religious convictions of the captain and his willingness to add the writing of sermons to his duties. Most captains quickly found a clergyman among the passengers to whom they could delegate.

Edward Towle, a single man travelling in second class, describes how the passengers gradually settled themselves into sociable groups and established themselves at home. Entertainment had to be supplied by the passengers themselves and keen amateur musicians were sought amongst them for concerts, together with ideas for debating societies and other ways of passing the time.

It would be surprising on a ship with so many relative strangers on board if there were not the occasional complaint, but by 1 September these complaints had increased in quantity and led to a meeting of 200 people to discuss them. In Towle's view, a few of the complaints were justified while others were 'frivolous and vexatious'. The complaints were presented to Mr Cox, the first officer, who managed the situation well and calmed the passengers. As the ship headed due south a crossing the equator ceremony was duly held with much horseplay and a few passenger grudges were played out.[32] The ship continued to steam directly down the west coast of Africa, but was meeting very strong headwinds, so by 14 September there were rumours that there were concerns about coal. Another diarist wrote about the captain's difficult decision.

At breakfast today the captain, who wore a very grave face and had been up all night, rose and addressed the passengers. 'Ladies and gentlemen, I think it right to inform you that from the continued state of the adverse winds it has been thought fit to return to St Helena for coals. I can't proceed to the Cape and we may have sufficient on board, but I have been always considered a prudent man and I hope with the large number of souls on board, you will still give me credit for the same. I have been misled as to the amount of coal on board. We may have sufficient to reach the Cape but considering the opposition which we have met with for the last two or three days from the continued headwinds, I much regret being obliged to adopt this course. It is I assure you a great trial to me.' In less than half an hour the vessel was turned round ... the course adopted seems by far the most prudent for supposing that we had arrived within 40 miles of the Cape and no more coal, there is every chance of her being wrecked.[33]

With 700 miles still to go to the Cape, Mathews, with some difficulty in the extremely strong headwinds, turned the vessel, stopped the engines and sailed 100 miles back to St Helena. It was an extremely rough and uncomfortable ride for all on board over three days. There were mixed views on who to blame. Towle decided 'the fault lay with the owners, Gibbs Bright and company by carrying an excess of merchandising and passengers which left a very insufficient room for coal.'[34] It would cost the owners more than just disgruntled passengers, since the ship had a contract to deliver the mails to Australia within 60 days and this is estimated to have cost them in the region of £3,000.[35]

Benjamin McFall at 28 years old was an experienced able seaman and had been at sea since 1840, most recently working on board Cunard's steamship *Britannia*. He was not impressed by Mathews.

Thursday 23, we arrived at St Helena and I suppose one of the dirtiest vessels that ever came in here. I expect this will be a very unfortunate affair for the owners. I am sure if they had a competent commander we should have been well on between the Cape and Australia. This one, I am sure he is not fit to be here. They say there is hardly any coals here. We cannot go without them. What we will do I cannot tell. I expect you will see strange accounts in the papers as the passengers will write to their friends and they are anything but pleased with the captain. They even say he cannot take us there and they posted a bill up offering five pounds reward to whoever could tell where the ship was. That was on Monday. So you can see what a comfortable ship I am in.[36]

Reginald Bright also noted that it was not a happy ship. Reginald was the son of Samuel Bright, the senior partner in the firm and was heading to Melbourne to establish another branch of the family firm as agents in that port. He kept detailed notes of the coal consumption, echoing the chief engineer's log.[37] They had been forced to purchase 100 tons at the exorbitant price of £5.17 shillings per ton and had also managed to persuade a naval ship, HMS *Penelope*, to part with a further 100 tons. 'We have been taking coals out of HMS *Penelope*. Here we are man of war's men, soldiers, and passengers working at the coals.'[38] The *Great Britain* left St Helena on 29 September with just sufficient coal to reach the Cape The ship finally arrived at the Cape on 10 October and remained there for seven days before setting off on its final leg to Melbourne.[39]

On the final part of the journey some passengers in first class began the almost routine task of organising a testimonial to the captain, but met with resistance in some quarters. 'Today I have spent some time canvassing the opinion of the after saloon passengers as to the propriety of addressing a testimonial to the captain on our leaving the ship. I am sorry to say I found a very bad feeling towards him, very many refused stating that he had been far from kind and courteous to them.'[40] A letter was carefully composed in 'a moderate style' and they managed 80 signatures with just ten refusals from the after saloon passengers. It was not the normal effusive praise.

> We the undersigned passengers in the after saloon of the screw steamship *Great Britain* having arrived at the conclusion of a voyage from England to Australia, beg, before we separate to express our regard and esteem for you as her commander. The proceeding direct from England to the Cape of Good Hope must be regarded as a great experiment in steam navigation, and under these circumstances, we believe you to have acted with praiseworthy discretion and judgement. Especially when placed in the very trying position of having to make a choice between two courses – that of putting back to St Helena or proceeding onwards against headwinds, with insufficiency of fuel. You decided on putting back – thereby showing a due regard for the comfort and perhaps safety of more than 700 souls. We beg to express our cordial thanks for all the care on our behalf during the voyage.[41]

She eventually arrived off Melbourne on 12 November and the whole voyage had lasted 83 days instead of the expected 60. Nevertheless, vast numbers of people came from all over Melbourne to view the ship and she caused quite a sensation. After five days at Melbourne she sailed on to Sydney with 300 passengers on board and was met by enthusiastic crowds and visited by politicians and leading members of the establishment, including the governor and his family. The Australian newspapers were full of praise. Just as New York had welcomed the arrival of the *Great Western* in 1838, the arrival of the *Great Britain* offered to cut journey times significantly and promised a regular fast connection back to Britain and speedier transmission of news. A local newspaper reported that there had been one man who had died, an elderly Chinese who was taken on board as crew in St Helena, but there was a second fatality, an able seaman who had fallen from the rigging.

The ship then returned to Melbourne and a grand ball was held three days after Christmas with invitations to some 400 local citizens. As usual she was then open to the public. She sailed at the beginning of January 1853 for Britain with 188 passengers and over 100,000 ounces of gold. She returned westward calling again at St Helena, the Azores and then Vigo in Spain before arriving in Liverpool in April to a warm welcome.[42]

The Steamer *Great Britain*. – The screw-steamship *Great Britain*, B. R. Mathews, Commander, whose arrival from Australia has been long anxiously looked for, arrived at Liverpool on Saturday. The pier-heads, for three or four miles length were closely packed with people. The *Great Britain* has brought passengers, and gold-dust of the value about £550,000, besides, a large amount in the hands of passengers. The passage of the *Great Britain* has been performed with an insufficient supply of coal throughout. At no time have more than four boilers been in use, in place of six, and yet, notwithstanding that disadvantage, she could, without the aid of canvas, steam upwards of 200 miles per day. She sailed from Melbourne on the 6th of January, and arrived, the 10th of February, at Simon's Bay, where she was detained upwards of eight days, coaling. She sailed from Table Bay on the 20th of February, and afterwards put into St. Michael's and Vigo, where she was detained four days. Her passage from Melbourne, including detention, is 86 days, but, deducting detention, only 72 days.[43]

Following the coal shortage problems the ship was put into dock for more alterations to resolve the lack of power. Alterations were made to her engines and her rigging so she became a conventional three-masted ship in appearance.[44] By August 1853 she was on her second voyage to Australia and this time she made Melbourne in 65 days carrying 34 first-class passengers, 119 second class and 161 third class. On her return she travelled eastward via Cape Horn, but had to shelter in the Falkland Islands for four days due to bad weather. She still managed to arrive in Liverpool after only 62 days. She carried 199 passengers, seven tons of gold, 23 bales of cotton, plus tin and 15 tons of mail.[45]

Mathews was still a divisive figure.

Some passengers are agitating a presentation to Captain Mathews of some plate for his services and more especially as it is probably his last voyage and it is thought that some parties meditate sending home a jaundiced account of him. Speaking from personal observation as I do, I hope my opinion may be useful in correcting any erroneous impressions that may have been formed. I have seen him repeatedly come to quell disturbances in the sleeping cabins from 10 o'clock till two in the morning. He has the good opinion of the sailors – he has himself tested our provisions and ordered better supplies and in the Lower Cabin he has been still more useful in preserving discipline. He said in our Saloon that if he had the power to prohibit it not a drop of spirits should be on-board except for medicinal purposes, and certainly spirits which are the root of all evils here. There are several things that I think might be better managed but they belong more to the Ship and Owners than to the Captain – I admit that at times he is crusty and disagreeable – but I am willing to make many allowances for him considering the vast responsibility resting upon him.[46]

Press reports, as ever, were positive, the *Illustrated Sydney News* wrote of the 'Queen of the Ocean' and her 'Commander' who 'has long since obtained a high position in the Mercantile Steam Marine; of his urbanity and kindness to all classes of passengers we need not speak; he is one of nature's proud nobility, and every inch a sailor'.[47] In England in the *Western Times* passenger William Poulton Green acknowledged 'the respectful treatment met with at your offices, the gentlemanly conduct of the captain (Mr Mathews) and officers, and the splendid accommodation that magnificent vessel, the *Great Britain*'.[48]

The complaints were not the first to be levelled at Mathews. As master of the *Great Western*, he had been the object of very similar complaints about a dirty ship, overcrowding, poor service and his manner towards passengers. Some found him a 'gentlemanly, courteous obliging little fellow' while others wrote, 'The present captain does not give as much satisfaction as did Captain Hosken; indeed, very serious complaints were made both respecting him and the steward.'[49] Mathews was generally a good sailor, although his performance on the first passage to Australia did not do him credit, but he was not suited to managing the complex needs of a passenger ship, as he tended to leave the hospitality department unsupervised and was brusque with passengers.

It had been rumoured that this second voyage back to Liverpool would be Mathews' final trip and indeed it was. John Gray, now first mate, took over. Mathews left the company's employment and took up a post as a Lloyd's agent in Melbourne, where his family joined him. He died there in 1869, aged 66, a highly respected member of the community.[50]

Mathews was not the only change. Reginald Bright had made several critical observations about the crew on his way out to

Australia on the first run in 1852. These seem to have resulted in action. He noted that there were insufficient stewards and the fore saloon was very dirty. John Nott, age 47, one of the senior stewards, was a 'good man but not up to an emergency, would do well with a well-trained staff under him', while 32-year-old John Clarke was 'too good for the bedroom'. Clarke left the ship on her return in 1853. In the galley John Birtles, age 33, was chief cook but was noted as 'very bad, not equal to his work'. He left the ship at the Cape of Good Hope on the first outward passage, while the second cook, 24-year-old Edward Blanchard, for whom this was his first ship, 'does all the cooking and his dishes are very much praised in the after saloon'. He left on return to Liverpool.

From the company's perspective the purser was a vital role, and in Reginald's view, Anjer was 'an honest man but very slow, not nearly clever enough to do his work, which is heavy, it requires a first-rate man'. He had joined on his first voyage in May 1852 and left in 1853. The storekeeper, Waddle, had a vital role, but he was judged 'very slow and thought to be a drinker'. He, too, left on return to Liverpool.

At the time of Reginald's voyage, John Gray was the second officer and he had several grumbles. 'Gray will not serve again as second mate, complains that no one backs him up, can't get through his work. Crew all taken up with passengers so no time to attend the rigging or sails, does not like looking after the second cabin passengers, they're always grumbling.'[51]

John Gray set out from Liverpool on his first voyage as master of a ship on 29 April, only to be involved in an unlucky incident. That night just off the coast of Anglesey a shock was felt on board with a loud noise and the engines stopped momentarily. Rapid examination found that the vessel was leaking from the stern near the propeller and so Gray returned promptly to port.

Once in dock it was discovered that she had apparently been in collision with a sunken wreck. There were evident marks of the collision on the keel and on the side but luckily the damage was slight and soon fixed, and she was again on her way.[52]

Captain Gray took the ship out to Melbourne in 58 days. The route now taken was more suited to a sailing ship with auxiliary steam. It took advantage of the winds across the Southern Atlantic heading west and then turned for the Cape of Good Hope. The sails were used as much as possible and steam when necessary. The testimonials to Gray were warm and appreciative. He clearly had a way with passengers, as compliments came not just from the saloon passengers but from the steerage passengers: 'A gentleman who is equally a gallant sailor and accomplished host.' They alluded to his 'fitness for the responsible post he holds, as shewn in the admirable discipline and order that had been preserved in every quarter of the ship', and congratulated him on having 'secured the services of officers who are worthy of acting under his orders'.[53] Other officers were singled out for praise, notably 'Mr Angus', who had saved the life of a crewman who had fallen overboard.

> None of us will easily forget the coolness and promptitude with which your orders were issued on the occasion of one of the sailors falling overboard, the happy rescue of whom was owing entirely to your energy and decision, and to the gallant conduct of your officer, Mr Angus.

A third officer was praised and that was Dr Alexander, the surgeon on the *Great Britain*, acknowledging 'that gentleman's great care and invaluable services'.[54]

Brunel's *Great Britain* and her master and crew were finally receiving the type of credit they all would wish for. It might seem that Brunel lost all interest in the ship after the great exertions of her rescue from Dundrum Bay and the subsequent sale to Liverpool. While he had no further official role as consulting engineer, he was kept informed about the ship by his loyal friend Captain Claxton.[55]

While no doubt of interest to Brunel, he now had other major shipping projects in hand. His sketch books show his continued thinking about ship design and the business world was eager to use him as their consultant engineer.

In the year following the sale of the *Great Britain*, a new company was registered, the Australian Royal Mail Steam Ship Company. The company's chairman was William Hawes, who was the younger brother of Benjamin Hawes, Brunel's close friend and brother-in-law.[56] The promoters of the company were a mix of merchants, a banker, Charles Barnett of Barnett Hoare & Co, Charles Walton, a shipowner, and Captain Frederick Hankey, RN.[57]

Their officers, solicitors, bankers etc., were registered and these included their consulting engineer, Isambard Kingdom Brunel. The company was established in January 1852 and by April 1852 it had obtained its charter and was awarded the government mail contract for Australia. They commissioned the building of four new steam ships overseen by their consulting engineer Brunel.[58] Two were built by Denny brothers in Dumbarton and the other two were to be built by John Scott Russell on the Thames. The role of the consulting engineer, Brunel, was to define the specifications of ship and engines. In order to fulfil the government mail contract speedily, the company bought an existing vessel. Built in 1849 by Scott of Greenock, the *Melbourne* was acquired by them from the Admiralty, quickly adapted and sent off.

The company was keen to increase its planned number of sailings. It had a contract for a bi-monthly service but wrote to the Colonial Office outlining the case for increasing it to a monthly mail supply, which from the company's perspective would bring faster financial returns.

Our company was formed in January last, our charter was signed in April and we have already one mail ship en route, another ready to start on 3 August, and two other vessels building to follow these, one in October and the other in December next, so that in less time than that in which many companies have started their first boat we shall have four – two of 1,400 tons and two of 2,000 on their way with the mails, passengers and freight to the colony and this has been accomplished in fulfilment of a contract taken on much lower terms than those of any previous or similar contract.[59]

This optimistic view of its future did not last long. The *Melbourne* set out on its first voyage and very rapidly ran into problems. On board was an Admiralty agent, Commander Philimore, as was required for a government mail ship. There were serious problems on the outward voyage and the vessel had to put into Lisbon to repair damage she had sustained. Here the Admiralty agent was in dispute with the master, who seemed unable to organise the repair of the vessel. Phillimore took over and wrote in November 1852 to condemn the master, 'who seems by his conduct to be wholly unfit to hold the responsible position of commanding a vessel of the Melbourne class'.[60]

The two vessels built by Scott Russell, *Adelaide* and *Victoria*, entered service in 1853. They were each 1130 tons with 450 hp engines, 260 feet long, plated in iron, screw-propelled and with four masts.[61]

The first to leave for Australia was the *Adelaide*, but she too experienced several difficulties for different reasons. She sailed from Plymouth on 3 January arrived at St Vincent on the 17th, then, after enduring a severe gale, stopped again at St Helena before reaching the Cape in March, where she remained for four days. The ship only used steam for a third of the time after leaving the Cape. Passengers were very dissatisfied with their long voyage and the chief engineer was sacked, with the third engineer promoted in his place.

The second ship the *Victoria* went out one year later and was much more successful, indeed it made very fast time, but by then the government mail contract had been cancelled.[62]

Brunel's first biographer Rolt asserts that Brunel had nothing to do with the design of these ships, which was left completely to Russell, but provides no evidence of this.[63] However, in Brunel's sketch book there is a sketch of a ship for the Australian Steam Navigation Company. He also made several sketches comparing four ship outlines, *Great Britain*, *Victoria*, *Wave Queen* and another unnamed vessel.

The *Wave Queen* was an experimental vessel sailing from Newhaven to Dieppe and intended to be part of a continuous communication linking with trains to Paris. It was trialled in

August 1852. The chairman, directors, and officers of the Paris, Rouen, and Dieppe Railway left Paris by an express train, and reached Dieppe in three hours and a half; and the same morning the chairman and officers of the London and Brighton Railway left London to meet in Newhaven. The vessel, which was to connect the two railways across the Channel, was the *Wave Queen*:

> For the purpose of rendering this as perfect as possible, a new description of vessel was tried, possessing qualities and arrangements of an unusual kind. Instead of the close cabins down in the hold of the ship, where passengers are usually confined, the vessel carries her first class passengers on deck in a large and luxurious saloon, surrounded with plate glass windows, where they are thoroughly protected from the wind and waves, and yet enjoy an extensive view, having abundance of light and the most perfect ventilation. Owing to the peculiar construction of the vessel the motion of the sea is much less felt than in steamers generally, a result which has been accomplished by making her of unusual length, so as prevent pitching. and by a proportion of beam and height which diminishes the rolling.[64]

Among the many dignitaries on board from the railways was one who was now very much Brunel's right-hand man, Captain Claxton, together with John Scott Russell.[65] Claxton's career might have ended with the Great Western Steam Ship Company to which he had devoted so much of his energy. He did appear to suffer a reverse of his fortunes briefly in 1849 as the company was suffering its most severe financial troubles and after the *Great Britain* had been rescued. Claxton's well-appointed house in Bristol was put up for sale with much of its contents. These included china and glass, a large brilliant cut-glass chandelier with twelve lights, a fine-toned grand piano, a circular library table, tables, beds and even curtains. All were sold by auction in what looked like a distress sale. From an elegant house in a good part of the city, Claxton moved with his wife and three of his daughters to a cottage in the tiny hamlet of Beachly by the ferry crossing to Wales.[66]

Claxton, the indefatigable fixer, had continued to work with Brunel and was involved with several of his engineering projects. In 1847 Claxton was working with the South Wales Railway to look at the best way of crossing the Irish channel. A steam ship was chartered, with Claxton in command, to sound the channel and make surveys. 'A better officer sent to serve the undertaking could not be found. We may rest assured that under his guidance the best and shortest route will be secured.'[67] In 1850 he and Brunel assisted Stephenson with the Britannia Bridge across the Menai Straits in Wales.

Via Claxton and others, Brunel was kept informed about the *Great Britain*. When the ship was docked in Liverpool in 1854 Claxton wrote with pride to Brunel on the condition of the bottom of the ship. 'There is not from stem to stern one single speck of rust.' A more detailed survey report was supplied: 'The *Great Britain* was not docked until Tuesday. I saw her in the water and out. I think she was very far from what I call foul from the light water line.'[68]

So while the *Great Britain* was making its mark in the emigrant trade with Australia, Brunel was similarly involved with the ships of a company trying to achieve the type of success on the Australia route that had been accomplished by Brunel's first two ships on the Atlantic.

However, by 1854 when the *Great Britain* had finally made a real success of the voyage to Australia, the Australian Steam Navigation Company had lost its mail contract and Brunel by now was interested in the Eastern Steam Navigation Company and had plans for a very large steamship that could carry its own coal all the way to Australia.

Strange rumours would continue to emerge concerning the *Great Britain*:

Steamer for the Pacific Trade. – It is reported that the steamship *Great Britain* is about to be purchased by a firm in Liverpool connected with the emigration trade, with the intention to fit her up for the conveyance of passengers on the Pacific, between Panama and San Francisco. £25,000 is mentioned as the sum demanded, and it is stated that the command will be given to an officer now engaged in the Halifax squadron. We are not able to guarantee these reports, but we believe some ground exists for them generally.[69]

It was an odd rumour and a tribute to her prestige that she was still much in the press, but in fact she was about to do something very different.

6

HER MAJESTY'S TROOPSHIP:
WAR SERVICE

In July 1854 the ship was now managed by the Liverpool and Australian Steam Navigation Company, a company established by Gibbs, Bright. But before she could go back to Australia, she was chartered by the government for war duty. In January 1854 French and British warships were sent to the Black Sea and later, naval squadrons were sent to the Baltic, the White Sea and the Pacific to blockade Russian naval bases. The campaign changed from an expedition to a major siege of the Russian fortress of Sevastopol in the Crimea.

The British and the French fleets had entered the Black Sea in January 1854 and the invasion was planned for that September. Large numbers of ships, both sailing ships and steamships, were under government contract. Mail ships, such as those in the service of P&O, Cunard and the Royal Mail Steam Packet Company, were commandeered as transport ships. The companies had little choice since it was written into their mail contracts. However, the payments were some compensation. These were not enough, so more ships, both sail and steam, were hired. The *Great Britain* was ideal as a troopship. [1]

To carry troops and horses to the Crimea the ship was stripped of much of her passenger accommodation. The situation in the Crimea was chaotic with poor administration of the large numbers of hurriedly commissioned transport ships, manned by crews more used to mercantile ways in normally well organised ports. The British army was ill-equipped and the success of the campaign relied on the efficient transport of troops and supplies to the Black Sea. The Transport Service of the Admiralty was responsible for this task but had not had to deal with anything on this scale since the Napoleonic wars, it was both bureaucratic and understaffed. Communications between London and the scattered British supply bases around Constantinople and the main Crimean base at Balaclava were poor. Balaclava harbour was a particular problem, being very small and overcrowded with few warehouses, dock labourers, unloading quays or wharves, or an efficient means of delivering supplies to the front. Added to this were the complications of Army command where the Ordnance Board was responsible for supplying guns and ammunition, while other parts of the Army had to approve shipboard accommodation.[2]

It was a far cry from the normally well-organised situation to which the Liverpool ships were used. Mackay was the owner of the Black Ball line of Liverpool and like Gibbs, Bright's Eagle Line, his ships had carried many emigrants to Australia. He wrote to *The Times* making a very unfavourable contrast between emigration agents in Liverpool and the situation in the Crimea.

Sir

The people of England owe *The Times* a very deep debt of gratitude for the bold and truthful exposure of the mismanagement of all departments connected with the war, but I think it also due to the public to know that there are men in the United Services who, when unfettered by the wretched staff system, which will, I trust, be soon rent asunder, can organise and get through an amazing mass of perplexing business, analogous to what is so horribly bungled in the Crimea.

In the first floor of the building near the Princes Dock (the rent of which cannot exceed £60 or £70) is the Government Emigration Office, and there, every morning before 10 o'clock, you will find Captain Schomburg, RN, surrounded by his officers, who, including medical inspectors and marine surveyors don't exceed a dozen people. In a few minutes afterwards they scatter for their day's work, and at 4 pm it is usually all over.

Now, mark what this knot of men get through in the year. They inspect the condition of every immigrant ship from the keel to the truck; order ventilation and light when required, and see that it is done; examine and taste every article of food that is provided for between 200,000 and 300,000 emigrants, of voyages from 70 to 140 days; see that the water casks are sufficient for their duty, and count every gallon of water required by law. Before sailing, two of them, one surgeon, pass in review on board the ship in the river every emigrant, and enquire each one if he has any complaint; has the detention money, if any, and which is checked from the embarkation ticket, instantly paid; makes the second survey of the between decks, berths, and hospitals, cooking houses, and then muster the officers and crew. In fact, for one half the salary of the first Lord of the Admiralty there is more work done than in the larger side of Somerset House.

'And how is this managed?' Is the natural enquiry. Simply because these immigration officers are permitted to exercise common sense in the conduct of their business, and, when thrown among mercantile men, can teach us our business in many ways. God forbid that government should export them to Balaclava, though there they would soon bring order out of confusion; but what I have stated I know to be true, I hope the day is not far distant when the public service of the country will in every department be equally well conducted, at the same time that the working bees will have the best share of the honey, instead of the drones, as heretofore.
yours faithfully TM Mackay Liverpool [3]

This was the chaotic situation that Captain Gray and his men had to negotiate. Also on board the *Great Britain* was 18-year-old Heywood Bright, son of Samuel Bright, who sent regular reports back to his father. In these reports he commented on the ship, the engines, the troops and relevant incidents. He was not very impressed by the soldiers of the British Army, calling them nasty, dirty and stupid, albeit admitting that some were militia. Large numbers of men were rushed into service to act as basic labourers. 'I was not struck with the British Army. The only thing they could do was eat all day long. The officers do not seem to care about the men.' Bright preferred the French troops who he said were 'always very polite and never grumble but appeared to be very glad to get their vitals'.[4]

Such a large ship like the *Great Britain* was a risk in the heavily overcrowded harbours used during the Crimean campaign. In Malta the propeller screw caught one of the buoy chains and snapped off one of the blades of the propeller, while in Kamiesh Bright reported that the ship carried away the bowsprit of one small vessel and the jibboom of another. 'I do not think Gray is to blame in any way as it is impossible to take a big ship like this in and out of harbours when they are crowded like a swarm of bees. At any rate we have done less damage both to ourselves and others than any other ship in the Transport Service.'[5]

The Crimean War was to set new standards in press reporting. Journalists such as William Howard Russell of *The Times*, and many others, provided grim and vivid descriptions of the appalling conditions in which the troops found themselves. The *Illustrated London News* gave this insight into the harbour of Balaclava where so many of the transport ships were destined.

Compared with the dull marshy solitude of the Camp, Balaclava is quite a metropolis; in fact, there's not another village in the world which, for its size, but show the same amount of business and excitement as is perpetually going forward in that little collection of huts which all the world is talking of under the name of Balaclava. The harbour is now like the basin of the London Docks, so crowded is it with shipping of all kinds; and from every one of these vessels, at all times a day, supplies are being constantly landed. Along a flat dirty causeway rather beneath the level of the harbour, are boats and barges of all kinds laden with biscuit, barrels of beef, pork, rum, bales of winter clothing, siege guns, boxes of ammunition, piles of shell, trusses of hay, and sacks of barley and potatoes, which are all landed in the wet and stacked in the mud. The motley crowd that is perpetually wading about among these piles of uneatable eatables is something beyond description. But very ragged, gaunt, hungry-looking men, with matted beards and moustaches, features grimed with dirt, and torn great coats stiff with successive layers of mud – these men, whose whole appearance speaks toil and suffering, and who instantly remind you of the very lowest and most impoverished class of Irish peasantry – are the picked soldiers from our different foot regiments, strong men selected to carry up provisions for the rest of the camp.[6]

In these conditions disease was rife. The P&O ship *Himalaya* on charter to the Admiralty left Southampton in July 1855 and was in Balaclava in September. The chief engineer noted in his diary the death of three men on board from cholera.[7] Mention of cholera on the *Great Britain* comes in a letter from Heywood Bright on 16 July: 'We are all well and the cholera seems to have left us altogether,' but he was over optimistic. In his final letter from Turkey dated 19 July he reported: 'We are to sail to day [...] We have had two more deaths from cholera, two stokers, the poor fellows got frightened, and the Dr says half killed themselves, but there have been no fresh attacks this last day or so, and the men seem in good health.'[8]

Cholera was a frightening disease, it was unpredictable and was thought to be linked to miasma, the general air. It killed rapidly and remorselessly; victims could fall into a coma within two hours and be dead within twelve. It was frequently carried by troops, being brought back from India early in the century. In 1854 French troops embarking for Gallipoli and Varna at Marseilles and Toulon carried cholera with them and ensured that a major outbreak occurred when they finally reached the theatre of war in Crimea. It is the only known time when the disease travelled the Mediterranean from West to East.[9] It is estimated that 95,000 allied soldiers out of the 155,000 who died in the Crimean conflict succumbed to various diseases.[10] At the time of this cholera outbreak, Dr John Snow would isolate a drinking water source in London as the cause of a local outbreak in the first study of epidemiology.[11]

Four men died from cholera on the *Great Britain*. The crewmen contracted the illness while off Scutari. The cholera bacillus spreads principally by contaminated water, and though the disease was present in both the Crimea and Turkey, it looks as if the crew were infected in Turkey, where it was reported that water could only be 'procured with difficulty'.[12] The first victim was John Simmons, age 53, an able seaman from Southampton, who died at 9 pm on 4 July. It was the task of James Beckett, the purser, to go through the dead man's effects and list them. The seaman's meagre property consisting of some clothing, hammock and pillow razors and four books was handed to the British consul. His next of kin was noted as Mrs Sarah Simmons of Toxteth in Liverpool. Four days later William Hellams died on 8 July and was buried at the naval burial ground at Therapia. The service was conducted by the Reverend Freeth, who was chaplain to the British military and naval hospital. Hellams was an assistant cook, aged 23, and his effects were similarly basic. The third man, Henry Abernathy, died on the morning of 17 July. He alone has a surviving tombstone, in the Haidar Pasha cemetery. The inscription reads: 'Sacred to the memory of "Henry Abernethy" stoker "Great Britain" steam ship who departed this life 17th July 1855 aged 59 years. May he rest in peace. His shipmates last token of friendship'. A Roman Catholic service was performed by the Reverend Malone. Abernathy's effects were very similar to the other two and he was also a reader, as he had three books. The last man to die was Patrick McGrady, age 20, who, like Abernethy, was a fireman. He died during the evening the same day but his remains were committed to the sea. A Roman Catholic service was performed, 'the surgeon judging it expedient that the body should be removed immediately for the general benefit of [the] ship'.[13]

On 19 July they left Scutari heading to Portsmouth and on board they had ten officers, 148 men, 'four women and one lady passenger'. The ship called at Malta and Gibraltar leaving on 25 July and 1 August respectively, and arrived at Spithead on 11 August; no further reports of cholera were reported during the voyage.

Heywood Bright, assisted the purser and represented his family's business. Such assignments were part of the training for future members of the family firm. Time was spent in the offices and in observing matters abroad. Having a member of the Gibbs, Bright family firm on board could be a mixed blessing but Gray was impressed by the young man and wrote warmly to Heywood's father: 'I shall regret when he leaves me to resume his duties at the desk for his kind attention and very gentlemanly deportment I really admire ... I think the voyage had improved him very much and, as you remark, he has seen little of the world and sea life and at the same time rendered his kind and efficient services in the stores department.' It certainly gave Heywood a set of very new and shocking experiences, witnessed war at close quarters. 'I never saw such a scene of desolation and riot.'[14]

Between 7 March 1855 and 7 January 1856 the *Great Britain* visited Marseille, Gibraltar, Malta, Genoa, Smyrna, Constantinople, Kamleash, Balaclava, Spitzia and Kertch and by the time her service finished in June 1856 she had carried nearly 45,000 troops. Brunel's *Great Britain* was joined in the Transport Service by the *Great Western* hired from the Royal Mail Steam Packet Company. In addition, Brunel's two most recent ships, the *Victoria* and the *Adelaide*, were also hired as transport, much to the relief of the Australian Steam Navigation Company, which had failed so signally to provide a steam mail service to Australia. The Crimea saw the end of the *Great Western*, which was broken up on her return, while the *Victoria* and *Adelaide* were sold. The *Great Britain*, however, still had many years of service ahead of her.[15]

It was during their time on war service in the Crimea that Captain Gray and the *Great Britain* were peripherally involved in a notorious and scandalous civil case in Britain and Ireland. In March 1855 one of the army officers who embarked at Malta was a Major Yelverton, who landed at Portsmouth in April 1855. The following month the ship headed back to the Crimea and stopping at Marseilles, she picked up a group of French nuns, the Sisters of Charity, and they were landed at Constantinople on 21 April. They were there to work in the hospital at Galata. One of their party, in uniform, but not a nun, was a Miss Maria Theresa Longworth. Captain Gray's letter confirming these details would later be published in the newspapers.

Great Britain, Dec. 7, 1859.

Dear Madam, – Replying to your letter, I beg that I have commanded the *Great Britain* steamship for the past five years, and perfectly remember Major Yelverton passenger from Malta to England in the *Great Britain* steamship. On referring to the log-books I find that Major Yelverton embarked at Malta on March 24, 1855, and landed at Portsmouth April 1855. Referring to the log-book I find that you embarked at Marseilles on the 12th May, 1855, and landed at Constantinople 21st of that month. Subsequent to your leaving the vessel I twice had the pleasure of seeing you at the Galata Hospital, on the last occasion in the dress of the French Sisters of Charity. Believe me to be yours respectfully (Signed) JOHN GRAY, Commander of the *Great Britain* [16]

The 19-year-old Maria Theresa Longworth first met William Charles Yelverton in 1852 on the crossing from Belgium to England. She was the youngest of six children of a soap manufacturer from Manchester and she had been largely educated in France and Italy following the death of her mother. Maria and William met again in 1855 during the Crimean War, Yelverton having returned to the Crimea after his passage home in the *Great Britain*.

Yelverton proposed marriage but his relatives apparently disapproved and it was broken off, but the couple stayed in touch. In 1857 they went through a private ceremony of marriage in Scotland when the couple read aloud a Church of England marriage service; Maria being persuaded that this was legal in Scotland as long as it was witnessed. Maria, being Roman Catholic, pressed for a church service and they were married again at the Roman Catholic chapel at Rostrevor in Ireland, but unknown to her this marriage was never registered.

In 1858 Yelverton married Emily Forbes, a widow and in 1859 Maria began many years of court cases to reassert her marriage claim. Maria was tenacious in her pursuit of Yelverton and appealed to many witnesses to support her case for, which she had a great deal of sympathy. It was in this regard that she wrote to Captain Gray in order to disprove statements made by Major Yelverton.

The case caused a sensation and there was daily coverage in the newspapers across Britain and Ireland when it came to a trial in Dublin. Both *The Times* and the *Manchester Guardian* sent their own correspondents to report on the trial. It lasted ten days, from 21 February to 4 March 1861, and Maria was the star, depicted as a young Roman Catholic of good family led astray by a 'member of the Irish landlord class', as Yelverton was heir to Viscount Avonmore. When the Major came to give his defence no ladies were permitted to enter the courtroom, since his defence included sexual details and the Major defended his seduction of Maria as she was 'not of gentle blood'.[17]

THE YELVERTON MARRIAGE CASE, Monday evening witnessed the close of this extraordinary trial in Dublin, which occupied ten days, and which excited an amount of interest rarely called forth by judicial proceedings. Mr Serjeant Armstrong, who commenced his address for the defence yesterday week, concluded on Saturday morning, when Mr Whiteside followed for the plaintiff. On Monday Chief Justice Monaghan summed up in an elaborate and nicely discriminative charge to the jury, which occupied nearly eight hours in the delivery. He left them to decide whether a marriage had been celebrated in Scotland or in Ireland, intimating that it was quite sufficient for them that a valid marriage had been celebrated either in Scotland or Ireland. The jury (composed of seven Protestants and five Roman Catholics), after a deliberation of about half an hour, found first for the validity of the Scotch marriage; secondly, for the validity of the Irish marriage; and, thirdly (in reply to the Chief Justice's question), that the defendant was a Roman Catholic at the time of the Irish marriage. The greatest enthusiasm prevailed through the city, and particularly in the neighbourhood of the courts. Mrs Yelverton was drawn in the carriage from the courts to the Gresham Hotel by the populace.[18]

Her triumph was short lived as it was ruled that the Dublin court was acting outside its jurisdiction. Maria continued to pursue the case and even had the help of Yelverton's uncle, who strongly disputed some of the Major's statements.[19] Seven court cases continued in London, Dublin and Scotland. With the help of friends, she eventually took it to the Appeal Court in London, where she lost. It became a famous legal case involving the jurisdiction of courts across Britain and Ireland, ecclesiastical law, civil law and the matter of a mixed marriage between Protestants and Roman Catholics. As a result of this case there was a Royal Commission into the laws of marriage, which published its findings in 1868 and several changes were made. It removed the impediment to Roman Catholic and Protestant marriages under civil law and by 1878 the irregular marriage system of Scotland was ruled out.[20]

Yelverton and Maria went their separate ways, Maria having spent much of her money on legal fees. She never remarried and supported herself by writing until her death in 1881.[21]

In 1857, after her service in the Crimea, the *Great Britain* changed shape yet again in a comprehensive refit. Gibbs, Bright had learnt considerably from the *Great Britain*'s previous voyages and this was all included in the changes. New lifting gear for the propeller was fitted by Thomas Vernon and son and changes were made to the decks, in particular to the deck house. She now had one funnel and three very much heavier masts and looked externally quite different from the ship launched by Brunel.[22] The refit took nearly nine months and at last, on February 1857, the newspapers could report the departure of the ship for Australia.

> Departure of the *Great Britain*. – The steamship *Great Britain* sailed from Liverpool on Monday for Melbourne, on her first voyage after repairs and alterations. Large numbers of people collected on the piers to witness her departure. She carries out about 520 passengers, and a heavy ship's mail. She is expected to make a very rapid run.[23]

The ship arrived in Melbourne following a good passage of 62 days; but her war service was not over. Severe rioting among Indian troops, which contemporaries termed the Indian Mutiny, broke out in 1857 and there was an urgent requirement for more troops to be sent. The *Great Britain* and the other large iron screw steamship of the time, *Himalaya*, were rushed out with troops for Bombay together with 68 other troopships. The *Great Britain* took men of the 8th Hussars and of the 17th Regiment of Lancers and made two voyages.

The ship left Liverpool on 24 September and headed for Cork, where there were several other transport ships getting ready to head for India. The Government inspectors checked the accommodation provisions 'with all of which they were highly pleased'. This is hardly surprising since she had only just been refitted for passengers. The steerage area was due to be home for troops and their horses. The surgeon on board, Samuel Archer, observed the baggage of the 8th Hussars arriving after lunch and 'the men themselves arrived in two batches; the dress is blue with yellow facings not unlike that of the artillery. The officers were mostly young men but some few had seen a good deal of service. When the Lancers arrive we shall be pretty well filled.'[24] The 8th Hussars were reported as 'having served with great distinction under General Greathed, its colonel, at Delhi, Agra, and elsewhere'.[25] The *Great Britain* headed for Bombay and arrived there in December. Archer was a young man intent on seeing the sights and he and some colleagues spent some days sightseeing and then returned to the ship to proceed to Liverpool. One passenger, handed to Archer for his professional care, became a great trial. Inverarity was a serious alcoholic with many delusions and managing his behaviour was a full-time occupation. Rather easier companions on board in first class were army officers and some wives and children who were returning. On the way back they stopped at the Cape, where further sightseeing occurred, and at last sailed into Liverpool.[26]

The next voyage after India was not to Australia but a visit to New York. The emigrant numbers for Australia were continuing to fall and gold fever was also well over now, which partly attributed to the decline. Migration to North America was still strong, so perhaps with another chance to take some of that trade she was well advertised.

The *Great Britain* Steamship for New York. – From our advertising column it will be perceived that the celebrated screw steamship *Great Britain*, Captain Gray, belonging to that well known firm Messrs. Bright, and Co., is to sail for New York on the 28th. This renowned and powerful steamship, under her well-known favourite commander, will be dispatched punctually on the appointed day, and, from her large size and excellent arrangements, offers more comfort to all classes passengers than any other ship afloat. Her saloon accommodations are of the very first order, and include baths, ladies' boudoir, &c. and the state-rooms are large and commodious.[27]

Her arrival in that great port was in August 1858. On board she carried 23 first class passengers and 62 in third class. The occupations of the third class were mainly farmers and miners. She returned to Liverpool on 9 September 1858 with 70 passengers, 60,000 dollars in specie and a full general cargo in twelve days. Following this Atlantic voyage, the *Great Britain* now headed back to Australia, advertising steerage passage at 14 guineas.[28] Then in July 1859 she made one final crossing to New York, perhaps in the hope that more business could be gained. It was not a success. She sailed on 1 July 1859 with just six passengers, three French, and three British. There was little interest in this ageing ship in New York. There was a brief reference in the *New York Times* on 11 July that the steamship *Great Britain* had left Liverpool and that 'she would probably bring accounts of the great battle of Solferino.'[29] There was more interest in Garibaldi's campaign to unite Italy. Keen to drum up business for the return voyage, the New York agent took out several advertisements.

Steam to Liverpool direct
The celebrated First Class screw steamship *Great Britain*, John Gray, Cmdr, will sale from Liverpool for New York, on Thursday, June 30, and will sale for Liverpool on return voyage on Thursday, July 28.
The state rooms and dining room for cabin passengers are in the spacious poop, and are fitted up with every convenience.
Rate of passage (first class only) $75, without wine and liquors, which are to be had on board.[30]

There was a direct reference to her absence from the New York route: 'As this ship has not been employed in the New York trade for some time shippers and others are referred to the following copy of circular from Liverpool underwriters.' There followed a brief statement dated in Liverpool on 23 June 1859, certifying that 'we consider her in every respect a first rate conveyance, and are prepared to insure goods in her on the very lowest terms.' It was signed by a list of Liverpool underwriters.

In the same edition, Cunard's British North America Royal Mail steamships were also advertising their ships from New York to Liverpool with first class fares at $130 and the second cabin cost equalling that advertised for first class on the *Great Britain* at $75. Gone were the days when Brunel's steamship could attract passengers paying the top prices. Cunard had the transatlantic business well under control, with a regular flow of steamships to Liverpool. Leaving from New York in July and August were the *Africa* and the *Persia*; while from Boston, Cunard despatched the *Arabia*, the *Canada* and the *Europa*.[31]

The *Great Britain* left New York on 28 July on her last run across the North Atlantic unheralded, except for a brief line in *Lloyd's List* on 11 August that notes her arrival with $300,000 of freight. One month later her creator, Isambard Kingdom Brunel, died at the age of 53 having at last seen the launch of his final massive project, the *Great Eastern*. This was planned to be the ship that would make the *Great Britain* obsolete by carrying vast numbers of passengers and all its own coal to Australia or India, obviating the need for calling at any ports on the way. This was the end of the connections between the *Great Britain* and her original building committee. Guppy was in Naples running his new business, Patterson had continued as a shipbuilder in Bristol, building both for private companies and the navy. Following the Crimean War he had to sell his yard where the *Great Western* was built. He later went back into business with his son in the old Great Western Company's dock from where the *Great Britain* was launched (and where it is today). He retired in 1865, moving to Liverpool where he died in 1869.[32]

Claxton remained active as an engineer. He filed patents for improvements in railway carriages, and by July 1864 was living in Brompton, London. He was the secretary of the Clifton Bridge Company, which was completed in 1864. On 7 March 1868, after seeing the successful opening of the Clifton Suspension Bridge in memory of his great friend, Brunel, Claxton died aged 78 in Chelsea.[33] Their ship sailed on, steadily continuing her run between Liverpool and Melbourne.

7

'THIS SPLENDID CRAFT': OWNERS, MASTERS AND CREW

After her war service and her farewell visit to New York, her owners now returned the *Great Britain* to the trade for which she was well equipped, the emigrant trade to Australia. From December 1859, until her return on her last passage in January 1876, she sailed steadily backwards and forwards between Liverpool and Australia. When Gibbs, Bright and Company of Liverpool purchased the *Great Britain* it appeared on the surface to be a complete change of ownership. She was registered on 17 April 1852 in the names of her new owners: George Gibbs, Robert Bright, Samuel Bright and Tyndall Bright, described as co-partners trading under the firm of Gibbs, Bright and Co.[1] Gibbs, Bright had been the Liverpool agents for the Great Western Steam Ship Company for some years, and the relationship between the two companies was in some ways highly typical of the interconnections between businesses in the shipping community during the early Victorian period.[2]

Gibbs, Bright had two main partners, George Gibbs and Robert Bright. Robert Bright came from a wealthy merchant family in Bristol and joined the firm of George Gibbs and sons in 1816. In 1818 the firm was renamed Gibbs, son and Bright, with a capital of £30,000. It had shipping, insurance and trading interests in Spain and Portugal and in the West Indies. Initially based in Bristol they had increasingly concentrated on Liverpool. This port had expanded with incredible rapidity and was considerably outpacing Bristol, not to mention the added attraction it offered of more attractive harbour fees compared to Bristol, and better access. While the office in Liverpool had initially been the junior partner, by the 1830s it was doing more business than Bristol. By 1839 Gibbs, Bright and Company's senior partners were Robert Bright as the managing partner, his brother Samuel Bright, Stephen Shute and George Gibbs. In the 1840s it expanded its trade with North America but by 1850 it was looking to Australia and the new opportunities opening there.[3]

Despite the growth of Liverpool, Gibbs, Bright continued its strong connection with Bristol. Robert Bright remained based there and was part of a wider group of businessmen trying to revitalise trade in the port. He was a shareholder in the Great Western Cotton Works, had a significant number of shares in the Great Western Railway and became its deputy chairman. He, together with Thomas Guppy, were two of the Great Western Railway directors who were early supporters of the Great Western Steam Ship Company. Robert Bright became an investor and director of the steam ship company.[4] On more than one occasion Bright had been a negotiator between the shareholders and the directors at crucial moments, as the *Great Britain* was being built and finances became ever tighter. After stepping down as a director of that company in 1843 owing to the death of his wife, Bright remained a senior partner in Gibbs, Bright.[5]

His older business partner, George Gibbs, was also Bristol-based and he, too, was on the Bristol committee of directors of the Great Western Railway company, investing £14,000. He was a member of the highly influential Society of Merchant Venturers in Bristol and a partner in the firm of Maze, Ames, Bush and Company who established the Great Western Cotton factory. Maze and Bush were also directors of the Great Western Steam Ship Company.[6] It was a close-knit network of overlapping business interests, typical of the time.

Gibbs retired from day-to-day management of the firm of Gibbs, Bright and Company in 1839 and from then it was dominated by members of the Bright family, Samuel Bright became the resident managing partner in Liverpool.[7] However, George Gibbs' name appears regularly as one of the two auditors of the Great Western Steam Ship Company.[8] So the acquisition of the *Great Britain* was an extension of the interests of Gibbs and Bright. They knew the ship well and they knew its potential. They were also more than well informed about the dire financial situation of the Great Western Steam Ship Company and in a position to buy the ship for a bargain price.

Gibbs, Bright's involvement in shipping to Australia began in 1850. A new line of Australian packets was advertised in April 1850 with the announcement of the launch of the *Albatross*, 1,026 tons register, a new ship. 'Sails remarkably fast and will be despatched on 20 April for Adelaide, Port Philip and Sydney, New South Wales. Her accommodations are superior to anything yet dispatched from this port.' Application was to be made to Gibbs, Bright & Co.[9] This was followed in June that year when the line was advertised as the Eagle Line, with three new ships, *Salacia* and *Petrel*, both 1,200 tons and the 1,500-ton *Condor*.[10] So Gibbs, Bright were already established with the emigrant trade to Australia at the time of the gold rush. They also promoted the qualities of their A1 fast sailing clipper ships and the passenger accommodation 'is of that roomy and comfortable description for which our ships are noted'.[11] There was another competitive advantage.

> We may also add that Messrs. Gibbs, Bright, and Co. land their passengers free of expense at Melbourne, which is not the case with other ships in general, thus saving much inconvenience and expense to the passengers, who frequently have to pay exorbitant rates to get landed with their luggage and other effects.[12]

Their line was known as the Eagle Line; the other important lines in Liverpool were the White Star Line and MacKay's Black Ball Line. Gibbs, Bright had a close working association with the latter.[13]

Until now the partnership had managed to raise the necessary finance to buy more ships through their own funds or by entering into ownership deals, essentially using the maritime system of 64th shares. Under maritime law each ship could have 64 shares and each owner of the shares had a form of limited liability and each ship was managed as a separate business.[14] So, for instance, when buying the 1,028-ton *Falcon* in April 1852 it was registered to Gibbs, Bright with 36 shares, John Bradbury 16 shares, Harry Thorp 8 shares and Thomas Taylor 4 shares. The 'said ship to be sailed, superintended, managed, insured, repaired and altered entirely under the management and control of Gibbs, Bright & co. All accounts to each owner as soon after each voyage as practicable.' The additional clause in the agreement between the owners was that 'any of the owners wishing to sell or part with his share shall first submit it to the other shareholders at fair market price.'[15] This system worked well until increasing needs for finance with larger, more expensive ships and a desire to manage them together, rather than accounting for them as separate businesses, began to put this system under strain. Raising further funds for the partnership meant possibly including non-family partners, which was always a risk. Robert Bright and George Gibbs had seen the challenges of the unincorporated joint stock company model used by the Great Western Steam Ship Company. But now, mid century, there was a new vehicle for business expansion with lower risks for investors.

Gibbs, Bright took advantage of the new company laws and established a chartered limited liability joint stock company, the Liverpool and Australian Navigation Company with a capital of £400,000 in 1856. This allowed the addition of external capital and the liability of each shareholder was limited to the amount of his individual subscription. The directors included Samuel Bright as chairman, and Robert Bright, together with other Liverpool names. They announced that the company was formed 'some time since for the purpose of supplying the valuable and rapidly increasing colonies of Australia with a regular and speedy communication with the mother country'. Pointing out the failure of the steam lines, 'having failed completely from their entire reliance on steam' and the challenges facing clipper sailing vessels, 'which succeeded them, relying wholly on their sails, [having] met with but little better success', they promoted their auxiliary vessels as combining the best qualities of each. They were also keen to obtain a government mail contract as Australia had no reliable regular mail connection with Britain and this would help to subsidise the service.[16] They now had, beside the *Great Britain*, the *Royal Charter*, an iron screw propelled sailing ship with auxiliary engines built by Sandycroft Ironworks, River Dee, Wales. After the bankruptcy of the shipbuilders, the vessel was finished by Patterson and launched in 1855.

THE STEAMERS "*ROYAL CHARTER*" AND "*GREAT BRITAIN*." (From the *Northern Daily Times*, 29th March, 1859.) The Australian mail has brought news of the arrival at Melbourne of both the *Royal Charter* and *Great Britain*; the former having made the passage in 64 days, and the latter, including a detention of five days in the Bay of Biscay and St Vincent, in 65 days. This is another instance, and we think most conclusive, that such ships of large tonnage, and perfect sailing powers, combined with the use of moderate steam, are those best suited for the trade and prosperity of England's most distant colony – Australia. Again and again have these ships made the passage in nearly the same number of days, and, in comparison with the following list of sailing vessels their present performance shows clearly the advantages they place before both passengers and shippers of goods.[17]

The newspaper confidently reported that 'the *Royal Charter* will leave again for Melbourne on 15th of May, and will, no doubt, be that favourite which her successful voyages so fairly entitles her to claim'. But on her return to Britain she was wrecked in hurricane force winds off Anglesey with great loss of life, over 450 people died.[18] The loss of the *Royal Charter* was a severe blow to the company, but at least they still had the *Great Britain*. Captain Gray was now one of the most highly regarded masters, receiving testimonials from grateful passengers for his 'attention, courtesy, and the invariable kindness which has made our passage so pleasant. We have ever found you ready and anxious to oblige and even to anticipate our wants, and to contribute to our information and amusement.'[19]

The master of the *Great Britain* had a sizeable crew to manage in three departments: the deck crew who managed the sails and other aspects of the ship's operation, both at sea and in port, the engine room crew, and the hospitality department. The total could vary between 140 to 170 signed-on crew depending on the number of passengers carried. For example, in 1863 there were 774 passengers heading from Liverpool to Melbourne. Captain John Gray was assisted by his first mate, Gilbert Peterson, and second, third and fourth mates. There were three quartermasters, three boatswains, 65 able seamen, eight ordinary seamen, three apprentices and one boy, aged 14, from Jersey in the Channel Islands. In addition there was one sailmaker and two carpenters, one joiner and a blacksmith, who might work assisting the deck crew or the engine room crew, or any other part of the ship that required metalwork. In the engine room was first engineer Alexander Neadon, with second, third, fourth, and fifth engineers, plus 24 firemen and one trimmer.[20]

The purser was James Beckett and he had one clerk and three storekeepers. In charge of the stewards was Austin Joseph Unsworth, assisted by John Campbell. There were 26 in total and one stewardess, Austin's mother Ann Unsworth, and one lamp trimmer, Michael Thunder. There were eight cooks, two bakers, one bar keeper and three butchers. Finally, another very senior and important person on board was the surgeon, Andrew Alexander,

These numbers reflect those who signed on, but Australia was a popular destination and desertion was still common. On this particular voyage 22 able seaman, one quartermaster, one steward and one fireman deserted in Melbourne and it would only be with difficulty that these numbers could be made up.

Desertion had been an even bigger problem for masters of visiting ships during the gold rush. On 15 November 1852, Benjamin McFall, a boatswain on the *Great Britain* wrote to his wife from Melbourne about the money that could be made by finding gold or working ashore:

> This place beats all I've ever seen, gold is like dirt among them. There is a man here, who was a dock gateman on the Albert Dock [Liverpool] when we went on our trial trip, who has just come from the diggings, sent home £400... And is going back to the diggings next week and expects with his knowledge of the place to go home independent in three months...There is upwards of 200 ships here, nearly all without a soul except the captain and the mate and (many) are for sale to pay their harbour dues not being able to go to sea for want of men... The country is most beautiful and the climate the same as England. Sailors is in most demand here from £16 to £25 or sometimes £30 per month to go in the coasters.[21]

Desertion was such a problem that one master had his crew arrested for attempted desertion and kept them in jail until the sailing day.[22]

The *Great Britain* crew were mainly from Britain and Ireland with a significant number from Captain John Gray's Shetland Isles and of course, many from Liverpool and surrounding areas. The non-British were mainly from Norway and Sweden. Giuzeppi Ruggier was from Malta. Joe Rogers, as he was generally known, was a great favourite on board and frequently mentioned in passenger diaries. This was not just due to his job, which was that of a lamp trimmer responsible for ensuring that the candles used in the passenger areas were properly doused at night, fire at sea being a major hazard. The main reason for Joe Rogers being mentioned so often was that he was famous as one of the heroes of the ship wreck of the *Royal Charter*. He saved the lives of 29 people by swimming with a line to the shore.[23]

Crew members were contracted, 'signed on', for each round voyage so it was normal for them to work across a range of ships. Those who chose to remain with the crew of the *Great Britain* included John Campbell who was one of many on board who remained for several years. John Anderson was a quartermaster, handling steering and assisting with navigation. He joined in 1864 as an able seaman and was quickly promoted. He remained with the ship until 1876, leaving at the age of 55. In the engineering section, Noah Beamish joined age 34 from Wallasey, Liverpool, and moved from junior engineer to chief engineer by 1857, his wages rising from 11 guineas per month to 20 guineas. He left in 1861 age 42, but had been joined the year before in the engineering department by Abraham Beamish, age 27. James Beckett, purser, joined as steward from the *Troubador* of Liverpool in 1853 at the age of 21 and remained through Crimean service until 1865. James Clare joined as an assistant steward when he was age 30 on his first voyage on a ship in February 1857 and he remained for another 20 years. There are also family connections on board. Between 1857 and 1874 at least one, if not two, of the cooks had the surname Dowdall. There was John Dowdall, who was joined by Henry in 1867 and also on occasions by Robert Dowdall, all of them from Newry, Ireland.[24]

The storekeeper's role was to keep an eye on the extensive range of stores on the ship sand manage them throughout the voyage. George Orams, born in 1822, had originally worked as a bootmaker in Liverpool and joined the ship in 1852 as a storekeeper as it headed for New York. He is one of the few crew members to leave his recollections. In the 1860s he wrote 'I feel sorry to look at our large and comfortable saloon with so few passengers. Our good ship is getting behind the times. Folks think her too old now.'[25] He and his family later settled in in Dunedin, New Zealand, their passage there apparently sponsored by Captain Gray.[26]

Working as a stewardess was the first paid opportunity for women to go to sea. It was the increase in emigration and the need for female assistants that brought about this change, although early steamships advertised the presence of a stewardess for the encouragement and comfort of female passengers. On the *Great Britain*, the names of stewardesses can be identified for 33 of the voyages. For women, particularly widows who might be free of the constraints of family, this was a way to travel the world. Many just travelled for one passage, but three were notable for their long service and for some working on board in the hospitality department became a family occupation.

Ann Unsworth joined the ship in 1858, age 42. Born in Shifnall, Ann was the second wife of Augustine Unsworth, a fustian manufacturer. She was no stranger to travel as shortly after her marriage she and her husband left England for New York in 1834, where their first child was born around 1837. They returned to England and by 1844 Augustine had taken over the Cross Keys Inn at 44 Regent Street, Clarence Dock, Liverpool. This was nicely situated close to thousands of very thirsty dockworkers, mariners and passengers.

Cross Keys & Manchester Inn, 44, Regent-Street, Clarence Dock, Liverpool. Augustine Unsworth, from Manchester, begs leave most respectfully to intimate to his Friends and the Public that he has taken the above well-known Inn (many years occupied by Mrs Johnson); and that he has, at considerable expense, entirely re-furnished the house with new and appropriate furniture, beds, &c., and that it is now complete and comfortable in every particular. The stock of wines, spirits, ales etc has been selected with great care; and the charges, in every department, have been arranged so as to endeavour to meet the approval of every visitor who wishes comfort at a moderate remunerating price. N.B. Well-aired beds.[27]

There were, however, financial problems and Augustine was declared insolvent in 1852.[28] At the time it appears that Augustine was already at sea working as a steward and so it is Ann who ran the inn and brought up their three children, Mary Catherine, Augustin Joseph and Thomas Henry.[29] Augustine was still a ship's steward in 1853, but died later that year. His son, Augustin or Austin Joseph, also signed up as a steward from the age of 14.

Ann followed in the family tradition going to sea in 1858 and joined her son on the *Great Britain*. The inn remained with the licence still in Ann's name until 1867. Her daughter was by now married and in position to run the inn. So Ann, together with her son, travelled to Australia and they were joined by Thomas Henry who signed on for one voyage in 1865, age 18, as a storekeeper. This was Ann's last voyage and she died two years later. Austin remained on board becoming the purser in 1866, a role he kept until his last voyage in 1873.[30]

On Ann's death the contents of the inn were sold at auction and they make an impressive list. This was not a cheap and cheerful establishment.

Sale on Thursday next by order of the executors of the late Mrs Ann Unsworth, 69 Regent Road, Wellington Dock.

Genuine household furniture, chimney and peer glasses, cases of stuffed birds, some paintings, chandeliers, china, glass and plated articles

Mr J F Griffiths will sell by auction on Thursday next the 18th instant at 11 o'clock precisely on the premises of the house in Regent Road, the whole of the above named furniture and effects by order of the executors.

The sitting room and parlour contain a large mantelpiece glass in gilt frame, pier glass, set of 12 mahogany chairs (reclining and easy) ditto sofa with loose cushions, set of mahogany dining tables, circular, mahogany loo, and other description of tables, handsome sideboard and cabinet, Brussels carpets, hearth rugs, fenders, window curtains, some china ornamental articles, chandeliers et cetera

In the chamber are half tester and other bedsteads, feather beds and bedding, mahogany and painted chests of drawers, wash stands, tables, chairs, toilet wear, dressing glasses, carpets, fenders, fire irons, drugget, useful kitchen requisites et cetera

Will be on view tomorrow Wednesday when catalogues may be had on the premises of the office of the auctioneer 44 Church St.[31]

After Ann's lengthy service, her replacements lasted just one voyage each. Mary McLellan, age 36 from Glasgow had previously been a stewardess on the *Penguin*. Then there was Elizabeth Baird, age 40 who had previously served on *Asia*. Elizabeth, however, was also on board with a new girl, Margaret Lynch. Margaret was a widow age 35 from Dublin and on her first vessel. Margaret re-joined the ship in December 1866 and remained for the next nine voyages returning to Liverpool in 1872. Margaret was to become well known. 'Mrs Lynch, the stewardess, has been in the vessel several years, and is an institution.'[32] Her final voyage was not as the stewardess but as an emigrant heading for Victoria, arriving there in December 1873. Margaret Gillespie, age 41, originally from Paisley, Scotland, was on three voyages. Unusually, she joined from Melbourne. She was an experienced stewardess having worked regularly on board the *Hero* plying between Sydney and Melbourne. The *Hero* was owned and run by Bright Brothers of Melbourne.[33] She joined the *Great Britain* in June 1872 and remained to the end of 1876.

Austin Unsworth was not the only long-serving chief steward. Alongside him, and then stepping into his place, was John Campbell, born in 1837 in Rothesay, Scotland, one of nine children. Campbell left behind detailed logs describing daily life on board, from births and deaths to Sunday services. He joined in 1860 aged 24 and was chief steward at 28, managing the meals on board, the hospitality crew and keeping the ship clean and well maintained. His diary entries note his continuing concern and care for the wellbeing of passengers of every class.

Wednesday, 17th
Light winds and fine weather. Steam and sail. Distance at noon, 225 miles. Lat.28"40S. Long.25"38W. The mother and the little girl born yesterday both doing well, and also the brother who was so poorly. All the other delicate folks doing well. The weather getting cool and pleasant. Dancing on Deck last night. All duties in each department going on quietly.[34]

He became the purser when he was 38 but life at sea was losing its gloss. 'Thoughts of the comfort and privileges of home are much in my mind.' In 1876 age 40, after 16 years at sea, he left the *Great Britain*. He died in May 1897 at the age of 61.[35]

On an Australian emigrant ship the most important person next to the captain was arguably the surgeon. While on a short transatlantic voyage his duties may have not extended much beyond some assistance to seasick passengers and crew accidents, on a two-month voyage much could happen. Infectious illnesses could spread quickly and the surgeon had to be able to deal with every possible situation from childbirth, typhoid, broken limbs to toothache. This was also an era when long voyages were often recommended for consumptive patients.

Surgeons had a busy time on board and they were able to practise privately by attending on cabin passengers and they probably earned more and had more opportunities for earning than they might have done had they remained in Britain; they were certainly better paid than surgeons on ships crossing the Atlantic.[36] Surgeons on the *Great Britain* were paid a nominal one guinea per month topped up with the income they could make by charging cabin passengers.

Supervision involved overseeing the sanitary regime, the distribution and cooking of rations, and looking after the sick. In addition to that, they were in control of discipline and the moral tone on board. On assisted emigrant ships, those controlled by the government, the role of the surgeon superintendent was largely to keep control of the emigrants and this relied on his authority.[37] On the *Great Britain*, while not in general being an assisted emigrant ship, the surgeon was still an important figure of authority, together with the captain. It was not an easy or necessarily popular role. Not all surgeons were capable of wielding that authority through lack of experience, incompetence, laziness, personality clashes with the captain, crew or emigrants, drunkenness or their own illness.[38] One surgeon on the *Great Britain*, Dr Puddicombe, was accused by eight people of 'neglect and want of attention'. This was reported in the *Brisbane Courier* on 2nd January 1874. Unknown to that newspaper, Dr Puddicombe had been ill and had died the day before.[39]

Medical journals throughout the 19th century frequently commented on the work of maritime surgeons and there were debates as to whether the 'worst of the best of British doctors went to sea'. Those on land tended to look down on those at sea, while many maritime surgeons often complained about the attitudes of the masters and the officers of private ships. But that could work both ways. Young medical graduates might see this as a chance to see the world and tended not to stay in the role, while those experienced surgeons who stayed in the role for years were well respected. One correspondent argued that they had chosen an unambitious career, but there were advantages, which included good earnings.[40]

The authorities had long been concerned about mortality rates on emigrant ships to Australia, which was what had led to so much regulation. These regulations reduced deaths considerably, but on a long voyage and with a mix of ages and backgrounds, they were inevitable in the 19th century. Between 1852 and 1882 over 140 deaths were recorded on board the *Great Britain* and more than 50 causes of death were attributed. They ranged from anaemia to teething, and premature birth to falling from the rigging. They include virtually all the major illnesses and diseases such as measles, bronchitis, heart disease, typhoid, cholera, appendicitis, apoplexy, alcohol poisoning and asphyxia at birth.[41] One of the most frequent illnesses and cause of death was tuberculosis, with two brothers dying on one voyage within weeks of one another. Senior steward John Campbell noted: 'There are three or four young men and women among our passengers in very poor health. They are making the voyage as a last remedy for Consumption.'[42] Serious infectious diseases like smallpox would mean the ship was quarantined, as happened in Melbourne in 1854.[43]

Mentally ill or deranged passengers had to be dealt with or preferably noticed before the ship departed. John Campbell noted one disturbed passenger:

Wednesday 20: We are not going today. Blowing a gale of wind... The poor woman has been so wild that it has been decided to send her on shore. Medical Doctor from Holyhead consulting with the Ship's Doctor. There will be some trouble landing her here.

Thursday morning, 21: The weather still looks wild and blowing. Storm signal still up. Not going today. The woman who was sent on shore yesterday is causing some trouble. One of the passengers has stated that he knew her to be the wife of a Limerick Butter Merchant. He is telegraphed for today. The Purser and Manager are busy at Court today about it. Thursday night, after a deal of trouble, the woman is left with her husband. He shipped her off to get rid of her, for she has given him a great deal of trouble through drink. All are glad that she is not left with us.[44]

Andrew Alexander had previously been the surgeon on the Atlantic passenger ship *City of Glasgow*. He joined the *Great Britain* in May 1852 and remained on board until 1868, arriving back age 49 to Liverpool in May 1868. He later returned to Australia where he died in 1871.[45] He was well liked by both passengers and crew. On one voyage he was presented with a silver snuff box by the firemen for looking after them so well.[46] This was a significant and relatively expensive gift from these crewmen. A popular surgeon could improve his earnings from grateful passengers and Alexander seemed to excel at this.

Feb. 20. Mr Andrew Alexander, surgeon on board the *Great Britain*, was presented by the first saloon passengers with a gold watch as an acknowledgement of his valuable services during his passage out; and when she was within a few days' sail of Liverpool, on her return voyage, he was presented with a purse of £50, accompanied by a testimonial, signed by the first-cabin passengers, offering their sincere thanks to him for his unremitting attentions in the discharge of his medical duties; and expressing an opinion that it was through his skill that, under God, though there had been many cases of malignant fever on board during the voyage, not one had terminated fatally.[47]

Samuel Archer, was just 22 when he joined the ship as a surgeon. Before joining the *Great Britain* he served on the troop transport

City of Washington, carrying invalided French soldiers home from the Crimean War. Archer was fascinated by nature, and collected shells and samples and wrote detailed notes in his diary about animals, birds and fish. He first sailed from Liverpool to Melbourne and back, and then to Bombay during the ship's time as a troop carrier in 1858. As the ship's surgeon he rose early every morning, checked the health of the ship, and tended to the sickly patients.

> ... Every morning I go round the ship and see that the places are kept clean and that all is going well in a sanitary point of view, then after breakfast I make up medicines and see the sick, which occupies me until lunch, which we have at twelve. In the afternoon I am generally at liberty but am now and then called away to see some one sick. The evening I generally pass on deck or read a little.[48]

He describes extracting teeth and treating patients with lung complaints, keeping the ship's hospital clean by using sulphur and putting chloride of lime, a bleaching powder, on the floor. After leaving the *Great Britain* he joined the Army in 1858 as Assistant Surgeon in the 98th Regiment, and rose to the rank of Surgeon-Colonel.[49]

Thomas Morland Hocken was born in 1836 in Stamford, Rutland (now Lincolnshire), England. He was only on board for two voyages but made quite an impression. He was described as 'barely 5 ft in height, a neat, dapper little man with a short-clipped beard and dark, lively eyes. Bustling, energetic, intensely industrious, he had a winning personality and infectious enthusiasm.'[50] A devout Anglican, he was the unfortunate object of one passenger's infatuation.

> Sunday April 21st: This Miss Dyson is a strange person, and somehow she has continued to get more laughed at than anyone else on board. She is an oldish young lady who wears her hair in ringlets and has little black beady eyes that are always darting about in search of admiration.

She is a clever person in some ways but very foolish in others, and the most foolish thing she has done is to make a dead set at the Doctor, and to talk about him and the compliments he has paid her, till people are always making fun of her about him. She gets almonds wrapped in paper sent her with the 'Doctor's compliments', which presents he has never sent her… Miss Dyson is going out as a governess, but I should be very sorry to leave children in her care. I do not like her ways.[51]

Occupying a position that was part crew and part passenger were members of the Bright family, owners of the ship. Heywood Bright, as noted in the previous chapter, was present on board during the Crimean service, but he was not the only one. It had long been a tradition in merchant partnerships or family firms to use family members as representatives of the business abroad, to act as key decision makers on behalf of the business when communication was difficult and letters took months to arrive. Such roles gave younger members valuable business experience. Henry Bright, son of Samuel Bright, was on the first voyage of the *Great Britain* to New York in 1852 age 22.

Doing business with Australia was a particular challenge and communications were slow, taking months to arrive. Having agents acting on behalf of the firm in foreign ports assisted with managing ships, passenger bookings and return cargoes. With the promised growth of the colonies in Australia, Gibbs, Bright decided to set up their own agency in Melbourne. The two brothers Charles and Reginald went there to establish an office for the parent firm in Liverpool.[52] Reginald sailed on the *Great Britain*'s first voyage in 1852. From then on members of the Bright family, notably Charles Edward Bright, were often on board the ship, a presence that would have brought mixed blessings, but even they could not override the captain's rule of the ship while at sea.[53]

Following on from Barnard Mathews, whose difficulties were witnessed by Reginald Bright, was Captain John Gray, who became the ship's longest serving master. Gray was born on the most northern Shetland island, Unst, on 8 December 1819 and gained early experience on his father's fishing boat. He was mate of the *Sea King* out of Liverpool from September 1847 to June 1849, then master of the *Loodianah* until April 52.[54] Incidentally, the *Sea King* was one of the two packet vessels that returned some of the stranded passengers from Dundrum Bay in 1846. Coastal vessels did not require certificated masters, but Gray gained his masters certificate on 1 June 1852 just before joining the *Great Britain* as Second Officer for the first voyage to Australia. Following Captain Mathews' resignation in April 1854, Gray became Captain. Contemporary accounts depict John Gray as a tall, stout (17 stone), square-built man with a fair complexion and a commanding presence. He was described as being highly intelligent, determined and a courageous man.[55] He managed all aspects of the ship, the different departments, crew and passengers, with tact. The long service of many of the crew is testimony to his management. Passengers valued a man who could keep control of a large number of humans living in cramped confinement for over two months. Two diary extracts from passengers in 1868 and 1869 illustrate his ability to keep the peace.

Today the second mate insulted Mr Buckle, one of the second-class passengers. This gentleman complained to the captain about it in the after part of the day. The captain sent to Mr Buckle to come into his stateroom and had the mate there to make an apology to him and the Captain gave him to understand that he would not have any one of the passengers insulted by his Officers but at the same time we the passengers must obey the orders given by him to his Officers.[56]

The *Great Britain* was commanded by Captain Gray, a Shetland man, I think, and known to more than half the population of the colony – the commander of a popular ship, deservedly esteemed for his tact and judgement. He knew how to handle a ship, its crew, and his passengers; when to raise steam and when to rely upon Sail Power. All told, we were over 800 people on board ... and as was inevitable on a long voyage, often quarrelled and fell out, more especially the ladies, and when they did Captain Gray displayed great tact in handling the unruly crowd.[57]

The *Liverpool Mercury* described him as: 'A self-made man who never forgot the class from which he sprang, and was always kind and generous to those who served under him'. He was extremely popular with passengers and crew alike, and is described fondly in many journal entries. He was generous with his time, teaching some passengers about navigation and how the engine worked.[58]

Captain Gray took pride in his passage times and managed the combination of sail and steam according to the sea and weather conditions. In 1870 they made the passage out in a record 56 days.[59] But the return passage tended to be longer. In 1865, for instance, the ship encountered a large number of icebergs on her passage between New Zealand and Cape Horn. For the first nine days after leaving Port Philip on 16 March there were light winds so steam was used, then as the wind picked up the screw propeller was lifted and the engines stopped. The icebergs were seen on the 26th, sail was reduced and the ship proceeded under steam. For the next ten days the ship continued to pass icebergs and on 7 April she was surrounded by them in a heavy sea with gale force winds. This slowed her passage and she eventually rounded Cape Horn on 10 April, reaching Liverpool on 20 May but she still managed a healthy passage of 65 days.[60]

Gray's main career was spent as master of just one ship on one route. Popular with passengers, well known in Melbourne and in Liverpool and respected by his crew, many of whom as we have seen served with him for long periods. But in a tragic incident on board, he left a mystery that is still unresolved. The ship, under his command, left Melbourne on 23 October 1872. On the passage out to Melbourne, the ship had made its fastest run yet. One of the senior crew was John Campbell, the chief steward. Campbell had served with Gray since 1860. On Tuesday, 26 November 1872 in the South Atlantic Campbell confided his shocked thoughts to his diary:

A great fear has come to us all this morning. We can't find the captain all over the ship. One of the storm ports on the lower saloon found down this morning, and the bedroom steward screwed it up the last thing before going to bed. That he should take his life is the last of our thoughts, though he was unwell. All who saw him last night had little thought of this... My mind can hardly realise that him who had been to me (with all his faults) a good master for over 12 years, is gone... God help his family... This sad end is so unexpected. Sick we knew him to be, and downhearted, as it always was in sickness, but we never thought of this. He has to all appearance got out of bed in the middle watch, gone through the saloon to the lower saloon and unscrewed the port. There is a lamp hangs by it all night, and he has taken time to lift the lid and put out the light before he dropped through. Servant went with his tea as usual this morning... What a sad tale for home. What will the thousands say, that have travelled with him on this ship he has been so long associated with. The two names: the *Great Britain* and Captain Gray.[61]

The ship was searched in the faint hope of finding him on board, but there was nothing else they could do but continue their course for home. Peter Robertson, the first officer, now took command. He had been with the ship since joining as second mate in October 1870. They arrived in Liverpool on Christmas Eve where 'Mrs Gray, accompanied by her daughters, had gone down to the Prince's Landing Stage for the purpose of meeting her husband on his return, and the melancholy tidings of his loss was then rendered additionally distressing.'[62]

This mysterious death grew in the telling as the newspapers latched onto the smallest detail to enhance their stories. The *Liverpool Mercury* referred to a serious, but unnamed, ailment which he had been suffering for some time.

Last voyage but one he was so ill that he hesitated then to go to sea, and remarked to an intimate friend that he was tired of the sea and wanted to have a good rest on shore. In fact, so settled was his determination to retire from the command, of the vessel that it is said he had made all the necessary arrangements for doing so at the termination of the late voyage, and that before sailing from Liverpool he remarked that that would be his last voyage in the Britain. When the vessel was lying at Melbourne, he had a slight attack of intermittent fever, but speedily got better. After the vessel proceeded to sea he again became unwell, and consulted the doctor who was on board, saying to him in the course of conversation that he was afraid he should never reach Liverpool. On the morning of the 26th of last month the captain's servant went into the cabin to attend to Captain Gray, but did not find him there. He subsequently again went to the cabin to look for his master, but as he was still missing he gave the alarm to the officers, and a minute search of the ship was made. The servant said that the last time he saw Captain Gray was about midnight on the 25th, when he noticed him leaving his cabin.

On examining his cabin it was found that some papers of a private character had been torn up and it also appeared that one of the ports or "stern windows" of the after cabin-some distance from his sleeping room – was open, and it is conjectured that through this the unfortunate gentleman either dropped designedly or fell accidentally into the sea. The opening of the window, by whoever done, must have been a work of some difficulty. A wooden bar placed across it on the inside had to be removed, and a large screw had to be unfastened. A lamp, which hung near the window, had been displaced, and the light blown out; and it was remarked that the roof of the cabin near the porthole was smeared with the marks of greasy fingers, and similar marks were observable on the outside of the hole, as if made by the hands of a person who had been holding on to the outside, or who had clutched at it in falling out. It is said that a few days before the occurrence Captain Gray gave the purser of the ship some money and valuables to take charge of, expressing at the time a fear that he would not live to see England again. An explanation, which may help to throw light on the mysterious affair, has been sent to us by an intimate friend of the late Captain Gray. He says that the captain suffered severely from neuralgia in the head, and it is supposed that he went to the stern window for the purpose of getting relief from the fresh air, when in some way – owing, to a lurch of the vessel – he lost his balance and fell overboard.[63]

Another news report suggested he might have been sleep walking.[64] While an Australian newspaper was more direct and referred to the 'suicide of Captain Gray, by jumping overboard from his vessel'.[65]

At first it seemed incredible that the cheerful and hearty commander whose ability and watchful care infused such confidence amongst those on board his vessel, should have ended a long and prosperous career so lamentably; but to those of his friends who were best informed concerning the state of his health of late, the catastrophe was not so unaccountable as it appeared to the general public. Some two or three voyages ago Captain Gray had a severe attack of congestion of the liver, and has never been quite the same man since. He remained the same careful captain and genial companion. To his passengers he was as attentive as ever, but he had lost that flow of spirits which formerly made him the life of every party at which he was present, and which on board his own ship tended in no slight degree to enliven and while away the tedium of a sea voyage. It was noticed when Captain Gray was in Melbourne last that he was labouring under more than ordinary amount of depression, and it is said that he had been heard to express the belief that the voyage he was about to undertake would be the last he should make in the *Great Britain*.[66]

Gray had known only the *Great Britain* since 1852 and his fondness for her was considerable. He spared no effort to ensure his ship was always in perfect order and, to this end, he meticulously examined every part of her during his tours of inspection. He once told a passenger, 'I love every plank of her. I pat her sometimes and I've promised her a rest if she will only get us home in less than 70 days.'[67] But he had sadness in his life, one of his two sons, Robert, died when only 18 years old in 1866.

There was one additional factor, the ship was elderly for a vessel and she was being overtaken by a new and more efficient breed of steamship. Gray knew that his days on the *Great Britain* were numbered.

8

'QUEEN OF THE WATERS': PASSENGERS TO AUSTRALIA

The main migration to Australia and New Zealand prior to 1850 was Government-assisted via a bounty system. The peak years for Australian emigration were from 1852 to 1855 after the discovery of gold, thereafter the numbers began to drop.[1] Emigrants then had to be encouraged with assisted passages as the growing economy of Australia needed people. By the time the *Great Britain* began her regular voyages Australia had changed from a dependent colony to a thriving importer and exporter as land was developed, new settlements appeared and the population grew. 'Total exports to New South Wales increased from £800,000 in 1837 to a record £2 2 million in 1840. Wool imports into Britain jumped from about 10,000 bales in 1831 to over 30,000 in 1837 and to almost 50,000 in 1841.'[2] London merchants took a burgeoning interest, developing trading associations and increasing their capacity in their wool warehouses.[3]

Shipping expanded rapidly to meet demand. In 1841, 251 ships arrived in New South Wales, compared with just 56 four years earlier. The noted firm of Devitt and Moore founded their shipbroking firm in 1836 and established professional freight brokerage in Australia. Plenty of labour was needed to fuel the economic growth and this required assisted migration. The growth of Australia fuelled a seller's market. Shipowners and ship brokers saw the opportunities for regular outward cargo on their way to the key ports in Asia or the Pacific.[4]

The growing colony needed not just unskilled labour but a wide range of occupations. From those that gave their occupations on board the *Great Britain* there were many professional men. These included accountants, including Olcher Fedden from Bristol, architects, blacksmiths, cabinet makers, bakers, engineers, farmers, chemists and jewellers. There were also doctors and surgeons. Agricultural men were much sought after to manage the growing livestock and mining skills were an obvious need, but one wonders about the bird stuffer and the three butlers. A problem for the historian is the self-labelling of many men who classed themselves simply as gentlemen, of which there were over 500 who travelled in 3rd class.[5]

The passenger set was predominately young and male, as would be expected with so many migrants. The young men went out first and, if married and deciding to settle permanently, sent for their wives later. Thirty per cent of passengers over the age of 16 were women and their average age was 30. Most of them had professions such as seamstress, dressmaker, or governess. Agnes Broadbent, a 48-year-old second class passenger who travelled in 1853, is a little more unusual as her occupation was listed as an ironmonger. There were dressmakers, laundry maids, laundresses, nine housekeepers, housemaids, 34 ladies, 11 nurses, three nursery maids and 273 servants.[6]

Liverpool was favoured by the Emigration Commissioners as an effective departure point for emigrants and there was a good supply of ships. 'Such favourable conditions in both the Australian and American emigrant trades gave Liverpool by far the largest single share of the business in mid century; in the early 1850s, two in every three people emigrating from the entire United Kingdom left from the Mersey.'[7] The busiest year for Australian immigration was 1854 and in that year 127 ships carried 41,000 immigrants under the auspices of the Emigration Commissioners.[8]

Before 1850 the control on passenger ships had varied widely. With the introduction of government agents in major emigration ports, practices became standardised and were often modelled on one man's recommendations. Lieutenant Robert Low of Liverpool was the first of the government agents, appointed in 1833. 'High on his list was keeping the sexes apart. Indecency superseded disease concerns. Women, unmarried females and boys under 14 were kept aft in steerage and adult males forward separated by partitions.' Low regarded himself as the emigrants' advocate and 'he besieged the Under-Secretary with a daily flow of plans, diagrams, measurements and tables that would assure the health, morals and instruction of the emigrants aboard ship.' The Emigrant Acts between 1842 and 1855 incorporated most of Low's ideas.[9]

The Commissioners laid down comprehensive regulations affecting almost all aspects of life on board for emigrants, from the space in the accommodation deck, the amount of luggage per person, the maximum number of emigrants carried, restrictions on the cargoes that could be carried by the shipowner in the rest of the vessel and specifications for food and medicine. 'A surgeon had to be carried free of charge, and he – not the ship's master – had responsibility for emigrants.'[10]

One of the ways in which order was kept on board amongst so many third class passengers was to appoint matrons. The women appointed to the role were often middle, or lower-middle, single women or widows in reduced circumstances, or 'wives whose husbands had 'suffered reversals'. They received a free passage and a gratuity at the end of the journey if deemed to have done their job well, usually £5. The surgeon superintendents, or the Emigration Commission's agent at the quayside, selected volunteers from amongst the married women or from those who applied for the role. From the 1850s many were selected and trained by the British Ladies Female Emigration Society at the ports of departure.[11]

The British Ladies Female Emigration Society was affiliated with an evangelical arm of the Church of England, the Society for Promoting Christian Knowledge. The British Ladies Female Emigration Society's concern were mainly for single women travelling alone and were largely spiritual, moral, and religious.

The committee, made up of the great and the good, including evangelistic ladies of good birth, took their responsibilities seriously. Single women's accommodation on board ship was inspected, clothing was distributed and the women were provided with sewing kits and shirting materials to occupy them on the voyage, and to provide a small income from the sale of their needlework (including shirts, handkerchiefs, and so forth) on board or in the colonies. Reading material, religious tracts and improving literature, was also distributed. The British Ladies Female Emigration Society was highly active and well connected.[12]

The matrons, who were carefully selected for their Christian principles, were expected to 'exercise a moral influence over all in the ship'. They had to chaperone the single women, to supervise female occupations as a guard against idleness, and to give spiritual and practical guidance.[13]

But finding the suitable person to fulfil the role was not always possible. Some were refused their gratuity or dismissed during the passage by the surgeon. The Emigration Commissioners, who had overall responsibility for the appointments defended themselves:

While we express our regret at the repeated failure of the matrons appointed to our ships, we may be permitted, in explanation, to point out the peculiar and rare qualifications required to make a thoroughly efficient matron. She ought to be physically robust and active, of a decided character and firm bearing, and of a high moral and religious tone of mind. She must also be of about the same rank in life as the young women among whom she is. placed, yet sufficiently superior to them in education and acquirements to secure their respect. Each of these qualifications may of course be found in many persons; and we do not believe that any of our matrons have been deficient in all; but the combination is rare under any circumstances, and especially in the class from which we have to select; and some of the qualities most necessary, such as decision and firmness, are exactly those which women may never in their previous lives have been called on to exercise in any considerable degree, and the possession or want of which may be unknown.[14]

The matrons were supervised on board by the surgeon, but on occasions his authority was challenged by some strong-minded matrons. The commissioners therefore wrote a set of instructions to all matrons on their appointment including those trained by the British Ladies Female Emigration Society and placed particular emphasis on the role of the surgeon.

Colonial Land and Emigration Office
8 Park Street, Westminster
Madam,
Instructions to matrons on emigrant ship
In reference to your appointment as matron on board [—], I am directed by the Board to convey to you the following instructions for your guidance.

1. All the girls, and also all the unmarried females not having natural protectors on board, will be considered as specially under your charge, *subject* to *the directions* of *the surgeon-superintendent.*
2. You will inspect the children daily, to see that they are clean and neat. You will be also careful that their stock of clothing is washed *at proper times*, to *be appointed by the surgeon-superintendent*, and that it is kept in good order.
3. You will assist the religious instructor or teacher, where one has been appointed by the Commissioners, in teaching the girls. But should no religious instructor or teacher have been appointed by the Commissioners, you will establish a girl's school, to be held daily, weather permitting, *at such hours as may be settled by the surgeon-superintendent.* All girls up to 16 years of age, who are not too young to derive benefit from it, will be expected to attend the school. You will not, however, confine your teaching to the children, but give instruction to any of the adult female emigrants who may be willing to receive it; and you will

endeavour as far as possible to render the more advanced, whether among the children or the adults, available for teaching the rest.

4. A portion of the elementary books put on board will be *handed over for your use by the surgeon, according to his discretion.* And in order that the female emigrants may always have something to do, and a motive for doing it, a supply of materials for work, of which a list will be supplied to you, will also be shipped at [—]. These materials will be placed at your disposal, to distribute from time to time as you may think best.

5. You will see that the single women are in their proper sleeping apartments as soon as it is dark, and that no male person is then on any pretence admitted, except the surgeon in his professional capacity. In the event of *the surgeon's attendance being required, it will be your duty to be present with him.* The key of the apartment will be in your custody.

6. With a view to prevent any irregularity, you are authorised to muster all the unmarried young women at any hour you may deem necessary for this purpose; but you will be responsible for the exercise of this authority with discretion. In short, you will endeavour to promote, by every means in your power, good order, regularity, and cleanliness, amongst the girls and young women under your special charge, *assisting the surgeon-superintendent in any measures which he may think necessary for that* purpose, *but bearing in mind that his decision must be received as final on all matters relating to the management of the emigrants.*

In conclusion I am to state that, in addition to your passage, the Commissioners will recommend you to the Colonial Government for a gratuity of [—], provided that you discharge the duties of your office to the satisfaction of the Governor.
I am, madam,
Your obedient servant,
Secretary[15]

By the 1870s the system had been so well established that suitable women had the opportunity to work as 'superior matrons', to be paid £15 per voyage and be provided with a return passage, thus turning the role into a professional rather than temporary role. They were provided with better accommodation and were partly paid for by the colonial authorities.[16] The fortunate few might even earn a generous bonus in lieu of a return passage. The British Ladies Female Emigration Society continued to train matrons until 1888.[17]

Two hundred and nineteen women made their passage as matrons on board the *Great Britain* between 1859 and 1873. Most of them were listed as married, just three are listed as single. Their ages ranged from 17 to 63 (Margaret Scott from Scotland who travelled in 1864) and most were in their 30s or 40s. Ellen Smith age 32 travelled in 1861 and Susannah Entwhistle age 37 in July 1860. Beatrix Pringle Tully, nee Robson, age 57 was the widow of David Tully. She was from Scotland and travelled to Australia with her two youngest daughters, Jane and Beatrice, in 1862. Teresa Carroll from Ireland was age 33 in 1863 and was following other family members out to Australia. She had been widowed just before the voyage and gave birth to a girl during the voyage.[18]

Acting as a matron provided valuable financial assistance to hard pressed emigrants and gave them a useful position on board. Selected cases show how family strategies were used to spread the cost of the voyage. On 3 May 1861 Sophia Mayne aged 36 boarded the ship and was listed as a matron and noted as married. Also on board was the youngest matron, Clara Mayne age 17, and also married. Further investigation reveals an apparent coincidence of Maynes, as there is Richard Mayne age 40, a labourer and listed as a single man, and two young children. On arrival in Melbourne they are noted in three separate groups on the passenger lists, Richard on his own, Selina Mayne age 17 with two small children, Sarah age 11 and Selina age 10, and then Sophia.[19] They were in fact all from one family. Richard was not merely a humble labourer but was a journeyman carpenter from Watford and he was not single, he was married to Sophia and the other three other girls, including the apparently married Clara, were their children. In this way they were able to subsidise their passage.

The years 1861 to 1863 saw the most matrons on the ship, with around 30 carried on each passage to Melbourne. In 1864 there were 17, two of whom had the same name. Mrs Jane Morton age 62 and Miss Jane Morton age 29 were matrons who embarked at Liverpool in May. Mrs Morton was a schoolmistress from Dorset. She was a widow and with her husband David she had had two daughters, Amelia and Jane, both born in Quebec, Canada. By 1851 Mrs Morton had returned as a widow to live and work with her sister, Salome Morse, who had a school in Sampford Peverell. It was a good-sized school for girls, with 17 pupils boarding. By 1861, Salome has gone and Jane Morton was in charge of twelve pupils included two boarders from the West Indies. She employed one governess and a professor of music. Just two years later she and her younger daughter, Jane, were travelling third class as matrons, so with assisted passage, to Australia. Something dramatic seems to have happened to cause them to leave the school and emigrate.[20]

A third mother and daughter set of matrons travelled in July 1861, Elizabeth Craven age 43 and Susannah Craven age 24. By the time they arrived in Melbourne they had an addition, a child Elizabeth. An illegitimate child may have been the reason for the voyage. Escaping from comment and gossip was one of the factors that encouraged emigration and the possibilities of a new life. There was also another apparently well-established family who headed to Australia with help in the form of a free passage.

In June 1862, 45-year-old Ellen Waterhouse, with her children Alfred, age 10, and Emma, age 8, were escaping a family tragedy. Ellen's husband Ralph B Waterhouse was a plumber and the victualler of the Wellington Hotel in Widnes. They lived well, employing a servant and an apprentice plumber.[21]

Mysterious Disappearance of Tradesman at Runcorn.
Considerable anxiety prevails, consequent on the sudden disappearance of Mr Ralph Waterhouse, proprietor of the Wellington Hotel, Runcorn Gap, Widnes. It appears that the

missing gentleman left his house on Monday week, on business, and from what can be ascertained called on Mr Crossfield, of Warrington since which time he has not been heard of. His road home lay by the canal, and it was thought he might have fallen in; but though the canal has been dragged over and over again, no traces of him have been discovered. It is rumoured that he may have met with some foul play, for, being a man of steady character and in good circumstances, his friends cannot entertain the idea of his having committed suicide. The police have been communicated with, and a description of the missing gentleman has been published the Hue and Cry; so far without any satisfactory result. He is described as 36 years of age, about feet high, broad set, black hair, full whiskers and moustache, and beard, bald on the top of the head. Dress: suit of brown cloth mixture, small pattern, or narrow stripe, light coloured cloth cap, with tassel, blue striped shirt, and black silk tie. Had with him a silver lever watch and silver guard.[22]

Acting as a matron to get free passage to Australia for Ellen looks appropriate after such a family shock, although it was a rather speedy departure, just seven months after her husband's disappearance. Perhaps debts drove them out or just a new life beckoned. However, there is one strange addition to this tale. The 1881 census shows Ralph B Waterhouse, a widower, born in Bolton, living in Lancashire. His occupation is given as practical plumber and lead worker. With him is his son Alfred, now aged 30 in the same occupation. Another family strategy perhaps.

A government report in 1854 noted the rate of mortality on emigrant ships as 1.23% and the 'very healthiness of emigration is attributed to the surgeons on the government ships' and also attributed it to the 'discipline from the matrons appointed over the single women'.[23] Even when not acting as a a government emigrant ship, the *Great Britain*'s owners conformed to many of the regulations.

There were many societies that sprang up in England linked to emigration like the British Ladies Female Emigration Society, and while they were concerned with the morals of single women, the next society in question was concerned with that other Victorian problem, the surplus of women in society, and that Victorian symbol, the governess. The Female Middle-Class Emigration Society was established in London in 1862 and it believed that there were more opportunities for the many single middle-class women prepared to emigrate to the colonies.[24] There were few recognised occupations for respectable single middle-class women and these were limited to roles such as dressmaker, milliner, teacher or governess. Widows often had more opportunity if they could take over their husband's business.

The problem for governesses was that many of them were out of work once the children grew older or boys were sent to school.[25] The Female Middle-Class Emigration Society assistance was a loan of their passage money interest free for two years and four months. The society organised the sailings and in view of the governesses' status, they normally travelled second class and where possible two or more were sent together in the same ship. There was also a system of agents, mainly a network of those in charitable societies, who could assist the woman as she tried to establish herself.[26] The majority of women travelled to Melbourne. Musical skill was seen as an essential accomplishment for a governess and it was also 'a crucial form of private and public entertainment', although one emigrant thought that Australians were 'not very particular about their music so long as it is a noise'.[27] The *Great Britain* carried seven governesses including two sisters heading for Brisbane. There may have been many more unrecorded, such as Miss Dyson, who had become so infatuated with Doctor Hocken in 1861, and was noted by a fellow passenger as going out to be a governess.[28]

Unfortunately, the need for governesses had been overestimated by the society and, indeed, the Colonial Land and Immigration Commissioners in 1863 made the point in their report that Australian colonies were mainly interested in skilled agricultural labour and domestic servants rather governesses. Less than half of those supported by the society to work in Australia as governesses found employment.[29] Salaries were a problem and it was generally said that servants did better than governesses since the former were in greater demand.[30] Benjamin McFall, the boatswain on the *Great Britain*, had observed this in 1852: 'Carpenters, brickmakers, stone-masons and all the hand trades are also in great demand. Clerks, gentlemen and fine ladies will starve here, that I know.'[31]

In addition to the matrons on board ship to keep the moral tone, another shipboard position was that of constable. Recruited from the male married emigrants they could earn a gratuity of between £2 to £5 on disembarkation if they performed their duties satisfactorily. They were selected by the surgeon to assist him in maintaining cleanliness on board. They would oversee the mess rosters, distribute rations, water, tea and coffee and ensure all was cleaned up after the meal. Sanitary constables supervised other important cleaning tasks like the daily cleaning of the steerage decks, fumigation, disinfecting, cleaning and repairing the water closets. Bedding had to be regularly aired and kept dry. Other voluntary roles included male and female hospital assistants.[32]

Keeping order on board ship with hundreds of passengers cooped up in their quarters on a long voyage was a big challenge. Discipline had to be maintained and for severe disorder or violence emigrants could be temporarily put in irons, such as an emigrant on board the *Great Britain* in 1869.[33]

... Nearly the whole day it rained a nasty drizzle – enough to prevent our going on deck, everyone seemed miserable and unhappy; an Irishman this afternoon got intoxicated & the Dr speaking to him was grossly insulted. He immediately ordered him to be put in irons for 6 hours, till he became sober.[34]

Returning successful gold prospectors might be targeted by opportunistic thieves on board. Henry Styles, a passenger from Melbourne, was charged at Liverpool Police Court with robbing a fellow passenger. Styles was accused of stealing 22 English sovereigns, four colonial ones, and several nuggets, at sea, on 25 April from John Moore. Styles was suspected and searched by one of the officers of the ship, who found on him just the 'number of English sovereigns stolen, and between 70 and 80 colonial sovereigns'. He protested his innocence, but was committed for trial at the Liverpool borough sessions. Another passenger, John Little, was charged with being an accomplice, but the case against him was dismissed.[35] It is curious that there is no further mention of the trial. Stealing such an obvious amount on board a ship seems an incredibly risky venture as there was nowhere to evade detection.

There were distinctions between first, second and third-class passengers in terms of their recreational spaces, both internally and externally, and in their cabins and their food. Rachel Henning travelled on the *Great Britain* in February 1861. Aged 35 and single, this was her second voyage to Australia, where she had a brother and a sister. Rachel's previous passage had been in a sailing ship. She wrote home to her family describing the difference between sail and steam.

> The captain sits at the top of the dining-table, next to the mast. Mrs Bronchordt sits next to him at his right, and next to her I sit; then a Mr Brand, a Scotchman; the pretty German and her husband sit opposite, and the 'commercials' down the same side. I can tell you nothing about the inhabitants of the different cabins, of course, I know none of them. I rather like a stout good-natured woman, who inhabits, with her husband, the one opposite to ours; but she is not a lady. You cannot think how dirty everything gets; hands, clothes everything is black. The white in my dress is in a most disastrous state. I never saw such a dirty ship.[36]

One notable first class passenger was the novelist Antony Trollope, who left Liverpool for Melbourne in 1871 with his wife to visit their son Frederick, who had a sheep farm in New South Wales. Trollope had been commissioned to write a book that would be helpful to those who were thinking of emigrating to Australia. Trollope wrote in his autobiography:

> When making long journeys, I have always succeeded in getting a desk put up in my cabin, and this was done ready for me in the Great Britain, so that I could go to work the

day after we left Liverpool. This I did; and before I reached Melbourne I had finished a story called *Lady Anna*.

His resulting book was *Australia and New Zealand*, and provided a sympathetic account of life in the Antipodes. But it was his novel, *John Caldigate*, which raised the issue of the precarious nature of the single middle-class woman's position in society. Euphemia Smith, a widowed middle-class emigrant strives for independence and for the right to decide her own fate. She defies conventional roles and becomes a gold prospector. Australia provides her with new opportunities, unlike England, to which she eventually returns. Trollope suggests that there were dangerous consequences for single women who violated conventional gender and class roles.[37]

Younger sons of the gentry also saw emigration as an opportunity. Travelling to Australia in 1875 was the Honourable Charles James Holmes à Court with his wife and one-year-old son. Age 31, Charles, was a younger son of William Holmes à Court, 2nd Lord Heytesbury. He was one of 15 children and had been brought up at Heytesbury House in Wiltshire. Some of his other brothers emigrated to Canada and South Africa.[38] Earlier in 1852, Frances Digby Legard, the 19-year-old son of a Yorkshire baronet, travelled to Melbourne in a second-class cabin. He remained merely a matter of weeks before returning to Yorkshire. 'What other form of character building his family had intended to achieve in gold rush Victoria, clearly he decided very rapidly that he wanted no part of it.'[39] But it was also quite possible he was merely visiting out of interest.

Other less well-heeled passengers included Edmund Veness, age 34, his wife Elizabeth age 30 and his sister Sarah age 40, all of whom travelled out in April 1863. Edmund was a gardener and was at one time listed as a curator working in Queen Victoria's gardens. Edmund's wife Elizabeth died one year after their arrival in Victoria, but Edmund remained and bought 29 acres of land from the government in 1866, cleared it and established a successful fruit farm. Ten years later he remarried aged 48 and had five children.[40]

Edward Towle travelled second class in 1852 and he hoped to work in farming or go gold prospecting. In his diary he commented on his fellow passengers:

> There seems to be a great mixture of characters on board, men who had been gambling the night before now appeared at church with a most devotional demeanour, and others who appeared to be very steady and sedate never went to church at all... We have French, Germans, Poles, Jews, Italians, Scotch and Irish on board.

Allan Gilmour provides a rare glimpse of life as a steerage passenger. At the age of 17 he travelled with his father and brother in cramped conditions, with four people per cabin, and he also noted the considerable gambling on board with as much as £500 changing hands in one game.[41] Another aspect of steerage noted was the presence of rats, despite Brunel's initial belief that an iron ship might deter vermin.

With passengers of all backgrounds on board together for two months, relationships, both business and personal, were formed, but it was often hard to escape prying eyes. Charles Chomley was born in Ireland but had settled well in Australia having come out with his mother when very young. When he came of age at 21 he decided to travel back to England on the *Great Britain* and go to visit his birthplace. He travelled first class and enjoyed the social life on board, but he also found the gossiping frustrating: 'I walk with one young lady more than another simply because she can talk sense & make herself agreeable and it is reported that we are engaged.' Louise Buchan was returning home with her mother and seven siblings to Edinburgh. The family emigrated to Australia in the 1860s but were now returning after the death of her father. Age 15, she too noted the shipboard relationships and had a keen talent for observation. 'The saloon was full of passengers, but they only stared at each other for the first few days. Then they began to bow and grin, and at last became intimate friends; some of the young ladies and gentlemen were very intimate friends indeed.'[42]

Travelling to Australia was not always a one-way ticket. Passengers returned for a variety of reasons. Allan Gilmour's father had hopes of setting up a business in Melbourne but died shortly after arriving. As Allan and his brother were under age they had to return to Britain. In 1869 another third-class passenger was deeply regretting his rash decision. Daniel Higson had been persuaded by his friend John Standing to buy 18-guinea tickets to Australia. Higson left behind his girlfriend Esther in Manchester. Deeply homesick at the outset, Higson remained just two years in Australia before returning to marry Esther and remain in England.[43]

There were considerable numbers of Cornish miners. Their hard-rock mining skills were in demand across the world and many travelled to Australia. This was a group well used to travel. They would depart for a mine unaccompanied and send money back to their families in Cornwall. They might be absent for some years before returning or might eventually decide to settle and then send for their wife and children.[44] Some never returned. An on-board newspaper, *The Cabinet*, noted two deaths on the ship in 1861.

In the text accompanying our log we notice the deaths of two of our fellow-passengers, which occurred at different dates since we crossed the Line.

Richard Sampson, whose death first occurred, died suddenly on the deck on the morning of the 17th November. A shock of apoplexy, with which he twice before had been threatened in his life time, was the immediate cause, and, as usual in such cases, the fatal result was sudden in the extreme. Sampson, who was a steerage passenger, was accompanied by his wife and two sons, and was on his way to join other sons in Australia, who had previously emigrated from his native place, St Erith, in Cornwall.

Samual Trinbat was the other passenger, and he was also a native of Cornwall, the parish St Just. He was accompanied by a son, 11 years of age, who will be returned by the "Great Britain", Captain Gray having been appointed as trustee over the means left for him, and for his sister and mother at home. The deceased, who had been captain of mines, was a man worn out with previous experiences in Australia, at home, and during a residence of 15 years in South America.[45]

Thomasine Williams Cornish from Redruth in Cornwall and her tin miner husband were travelling to Melbourne in 1852 and her occupation is listed as a dressmaker which would have been a useful way of earning small amounts of additional income on board the ship. She was subsequently widowed twice and all her husbands were miners. Many of these Cornish men and women settled in places like Moonta in Australia establishing a Cornish enclave.[46]

The gold rush had caused great consternation in Australia for the authorities. The report to the Colonial Minister from one official referred to 'the calamity which, by the recent discovery of gold in New South Wales, has so suddenly fallen upon the fairest and most promising portion of our colonial possessions in Australia'. Four hundred men a week were arriving and many others were 'finding their way overland and some are stealing horses to find their way'. He estimated that 'Victoria is losing weekly as much of its most efficient male labour as is usually sent out in three emigrant ships, and as yet this tide is but settling in.'[47] Many hopeful fortune seekers arrived via the *Great Britain* and many were to be disappointed. The most successful person was William Poulsen Green from Hertfordshire. He had resigned from his job as railway clerk and bought a steerage ticket to Melbourne with the original idea of finding work as a clerk. A chance encounter with a friend persuaded him to try his luck in the goal diggings and he got very lucky indeed. Green discovered the largest nugget ever found, 40 lbs, which enabled him and his wife to return to England and subsequently emigrate to Canada.[48]

Passengers might also travel for sport, as was the case in 1861 when the *Great Britain* carried the first all-England team of cricketers to tour Australia. There were several cricket clubs in Melbourne who were keen to set up the tour and began to fund raise. After a couple of unsuccessful starts, two Melbourne-based businessmen, Felix Spiers and Christopher Pond, guaranteed the sum of £7,000. An agent was sent over to England to bring out a picked team of cricketers. Twelve cricketers were selected and signed up and received £150, first class passage (70 guineas) and all expenses. Heathfield Harman Stephenson of Surrey was the captain and other Surrey players were William Caffyn, William Mortlock, George Griffith, Tom Sewell and William Mudie. From Yorkshire came Edward Stephenson and Roger Iddison while Tom Hearne and Charles Lawrence represented Middlesex. Finally came George Wells and George Bennett from Kent. George Wells used the tour for a holiday and had left earlier with his wife for Australia. The plan was to return by May of the next year in time for the start of cricket season in England.[49]

The ship finally anchored off Sandridge Pier, Melbourne, on 23 December 1861 where great crowds had gathered to greet the ship, around 10,000 people. The team played 15 matches beginning on 1 January 1862. Attendances were extremely good and the fifth match at in Sydney had an estimated 15 to 20,000 spectators. The England team won six games, drew four and lost two on rough grounds, some of which had been hurriedly created for the occasion. Such was the success that Spears and Pond 'offered the players £1,200 to stay on for a further month'. But most returned home.[50] It was the first All-England cricket tour, the first commercially sponsored and the first time the term 'test match' was used.[51]

In 1863 the second all-England tour of Australia and New Zealand travelled on the *Great Britain*. Melbourne Cricket club invited Nottingham player George Parr to select and captain a team. He managed to persuade E. M. Grace, brother of W. G. Grace, to tour for the first time. Surrey player William Caffyn was the only one who had been on the first tour. The group was made up of Grace from Gloucestershire, George Parr, John 'Foghorn' Jackson, R. C. Tinley and A. Clarke, all from Nottinghamshire, William Caffyn, Julius Caesar and

Tom Lockyer all from Surrey, Tom Hayward, George Tarrant and Robert Carpenter from Cambridgeshire, and George Anderson representing Yorkshire.[52]

Born in Downend, Gloucestershire, Edward Grace was one of a trio of sporting brothers. His father was a doctor and very influential in setting up Gloucestershire County Cricket Club. With the English cricket team, Edward toured America before being selected for the Australia and New Zealand tour of 1863. His on-board diary includes the following comment: 'I have not been introduced to any of the ladies yet for a very good reason as there is not one sufficiently to my liking.' Keeping occupied on such a long journey was largely a matter for the passengers to arrange. The cricket team kept themselves fit with various sports on deck when the weather was right. Grace also talks of other entertainments such as concerts, singing, dancing, playing cards and sweepstakes.[53]

During some voyages the passengers would create their own newspaper and on the first cricket tour that newspaper was called 'The Cabinet: a repository of facts, figures, and fancies relating to the voyage of the 'Great Britain" S.S. from Liverpool to Melbourne with The Eleven of All England and other distinguished passengers.' It was subsequently published in Melbourne by its editor Reid, one of the few editors of a *Great Britain* shipboard newspaper who was a professional journalist. Alex Reid was from Wick in Scotland, a bright boy who did well at school and then went on to the University of Edinburgh. He went into journalism working on several regional newspapers in England. He travelled to the United States and Canada and became a journalist on the *Toronto Globe* but his health was not strong so he returned to Scotland. He still had the urge to travel, so he sailed with the *Great Britain* in 1861 to Melbourne. During his eight-week voyage, he edited and wrote most of *The Cabinet*, for weekly circulation among 600 passengers and crew. Passengers then raised £30 to have it printed in Melbourne. Its success found him a job on the *Dunedin Times* in New Zealand, where he remained for 16 years. He died in a drowning accident.[54]

1 *Above left*: Isambard Kingdom Brunel. (Clive Richards Collection)

2 *Above right*: Thomas Guppy. (*A Short History of the Great Western* published 1938)

3 *Above left*: William Patterson. (*A Short History of the Great Western* published 1938)

4 *Above right*: Captain Hosken portrait as naval officer. (Hosken family/SS Great Britain Trust)

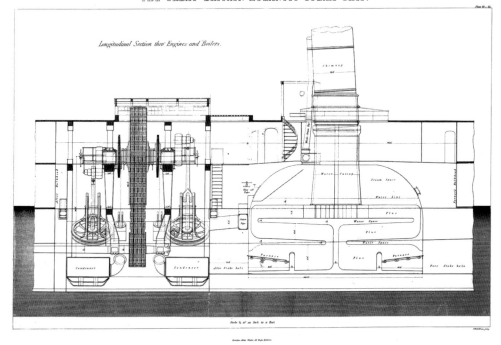

Longitudinal Section thro' Engines and Boilers.

5 1843 engine and boilers. (SS Great Britain Trust)

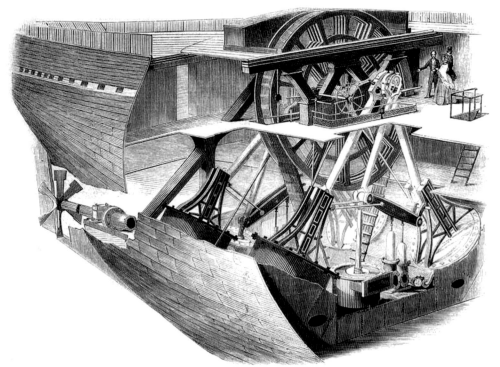

MACHINERY OF THE GREAT BRITAIN STEAM-SHIP.

6 Drawing of drive shaft and propeller. (SS Great Britain Trust)

7 Replica propeller on the *Great Britain*. (SS Great Britain Trust)

8 Crowds at the floating out in the presence of Prince Albert 1843. (SS Great Britain Trust)

9 The *Great Britain* launched into the floating dock. (SS Great Britain Trust)

10 *Great Britain* at last on its way down the Avon by Joseph Walter. (SS Great Britain Trust)

11 Passengers in the first class dining saloon. (SS Great Britain Trust)

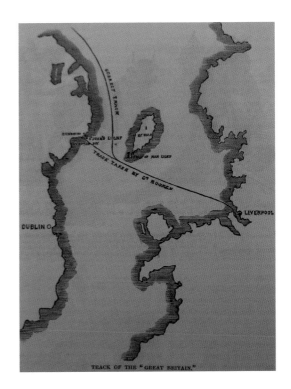

12 The fateful track to Dundrum Bay. (*Illustrated London News*)

THE "GREAT BRITAIN" STEAM-SHIP.

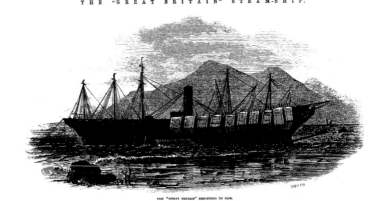

THE "GREAT BRITAIN" BEGINNING TO SINK.

13 The salvage operation. (*Illustrated London News*)

14 *Great Britain* grounded at Dundrum. (*Illustrated London News*)

15 Britannia Bridge Conference 1850, Claxton standing third from right, Brunel seated far right. (Institution of Civil Engineers).

16 Barnard Mathews, his officers, cadets and Francis Pettit Smith, patent holder of the screw propeller on *Great Britain*. (Clive Richards Collection)

17 *Above*: Painting of the *Great Britain* in Melbourne harbour. (SS Great Britain Trust)

18 *Right*: Captain Gray, long-serving captain of *Great Britain*. (SS Great Britain Trust)

19 The small harbour at Balaclava. (*Illustrated London News*)

20 *Great Britain* in Table Bay, 1852, by John A. Wilson. (SS Great Britain Trust)

21 Second class passenger ticket. (SS Great Britain Trust)

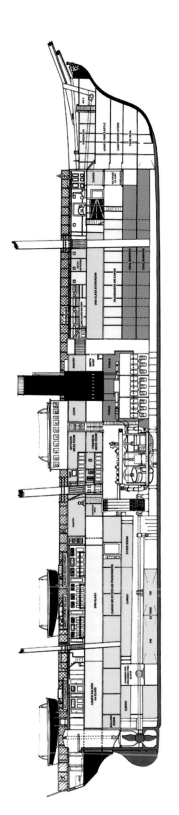

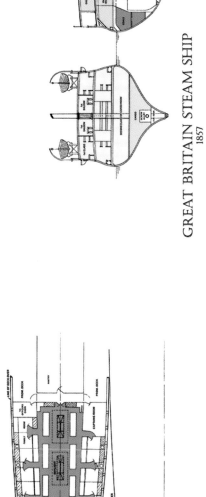

GREAT BRITAIN STEAM SHIP
1857

22 The layout of *Great Britain* in 1857. (SS Great Britain Trust)

23 *Above left*: Live poultry on the deck of the *Great Britain*. (SS Great Britain Trust)

24 *Above right*: Watercolour of *Great Britain* by a passenger in 1873. (SS Great Britain Trust)

25 *Great Britain* in 1872. (SS Great Britain Trust)

26 Poster advertising passage on the *Great Britain* in 1873. (SS Great Britain Trust)

The Historical old "Great Britain" now used as a Wool
Store Hulk by the Falkland Islands Company, Ltd., Stanley, Falkland Islands

27 *Above*: The
Great Britain as a
wool store in Port
Stanley, Falkland
Islands. (SS Great
Britain Trust)

28 *Right*: The draft
of Corlett's letter
to *The Times*.
(SS Great Britain
Trust)

FIRST IRON STEAMSHIP

From Mr. E. C. B. Corlett

Sir,-The first iron built ocean-going steamship and the
first such ship to be driven entirely by a propeller was
the Great Britain, designed and launched by Isambard
Kingdom Brunel. This, the forefather of all modern ships,
is lying a beached hulk in the Falkland Islands at this
moment.

The Cutty Sark has righly been preserved at Greenwich and
H.M.S. Victory at Portsmouth. Historically the Great Britain
has an equal claim to fame and yet nothing has been done to
document the hulk, let alone recover it and preserve it for
record.

May I make a plea that the authorities should at least
document, photograph, and fully record this wreck and at
best do something to recover the ship and place her on
display as one of the very few really historic ships still
in existence.

Yours faithfully,

E. C. B. CORLETT.

The Coach House, Worting Park,
   Basingstoke, Hampshire, Nov. 8

Times Nov. 67.

29 Salvage crew welding inside *Great Britain*. (SS Great Britain Trust)

30 Crew relaxing in a temporary swimming pool on the deck of *Varius II*. (SS Great Britain Trust).

31 Lord Strathcona, representative of the Great Britain Project, making a speech in Port Stanley. (SS Great Britain Trust)

32 The *Great Britain* escorted on her transatlantic return passage. (SS Great Britain Trust)

33 Richard Goold-Adams meets Jack Hayward at Lulsgate Airport. (SS Great Britain Trust)

34 Towing the *Great Britain* round the Avon's Horseshoe Bend in 1970. (SS Great Britain Trust)

35 L-R: Ewan Corlett, Jack Hayward and Lord Strathcona with the ship's hull in the background. (SS Great Britain Trust)

36 Safely back in her original dock. (SS Great Britain Trust)

37 From the bows.

38 Floating in her glass sea. (SS Great Britain Trust)

39 *Great Britain* lit up at night. (SS Great Britain Trust)

The onboard newspapers varied in quality. *The Albatross* in 1862 was a mixture of poetry, news, quizzes and articles and edited by Colonel Sir James Alexander. It included conundrums – 'When was the walking stick invented?' – and advertisements – 'Wanted as soon as possible, the trade winds. Supply to Captain Gray who hereby offers a reward of a quick passage to the producer.'[55] But not all newspapers gained approval. 'They have tried to start a paper on board the "Great Britain" but Dr Puddicombe who is also a magistrate forbids it, said it became too personal sometimes & gave offence so a stop was put to our proposed amusement.'[56]

On occasion there were professional musicians on board. Howard's Serenaders were a group of five men who sang popular American songs. They joined in Melbourne for the short trip to Sydney.

HOWARD'S SERENADERS

FIVE in number – the only GENUINE Company of ETHIOPIANS in the Colony. Another Grand Performance takes place TO-MORROW, Friday, March 18th, at the Royal Hotel. N.H. On Wednesday Evening, March 30th the Entertainments by the above inimitable performers will be under the distinguished patronage of His EXCELLENCY the GOVERNOR-GENERAL, SIR CHARLES A. FITZROY.[57]

Some passengers were simply tourists with the time and money to indulge in travel.

Return of Mr Isaac Pierce from Australia – Towards the close of last year we reported an interesting farewell dinner, which took place at the Old Swan Inn, given to Mr Isaac Pierce, who was about to leave his native town for a time on a visit to his brother, Thomas Pierce, in Australia. Mr Pierce had a beautiful run out in the *Great Britain* steamship, and about a month's ramble in the country in company with his brother, then took a fancy to return with the same vessel that carried him out so safely, and on Tuesday last landed in Liverpool again.[58]

In 1866 Mary Crompton was on board the ship and began her journal with these words: 'I do not mean to keep a regular journal, for I think the ship journal is a very disagreeable thing to keep and very tiresome to read afterwards.' She had been in Australia from the age of four and in 1866 she married Joseph Crompton, who emigrated from Liverpool in 1860 on board the *Great Britain*. They were now voyaging back to Liverpool together on their honeymoon. She drew a delightful little sketch of the cabin. They were first class passengers and she felt highly honoured by the captain 'putting us at his table'.[59]

Other officers, including the surgeon, were expected to host dinner tables when duty allowed. The young surgeon, Archer, confided his thoughts to his diary. 'I had the honour of having the Dean of Melbourne on my right whose son is going home with us and also Mr Jones of the *Argus*. The generality of the passengers seem to be self-made men and there are very few females, still fewer ladies.'[60]

Early assessments were made of their fellow passengers. Mary Crompton described the saloon with six tables all covered with red cloths, long seats on each side and her main complaint was that it was very draughty. She assessed her fellow passengers. The man sitting next at the table was Mr Fenwick and she decided that he seemed disposed to be a 'pleasant, grave, gentlemanly man' with a pretty wife. While Miss Farrar was promising, 'Miss Farrar I think I should like very much, she is plain and ladylike, seems very sensible and as if she worked very hard at home.' 'There are swarms of children, I don't know who they may belong to, 24 in all.' The weather was quite rough and many of the passengers were seasick; on 18 May she was the only lady at breakfast and there were just two of them at lunchtime. 'They keep a magnificent table on board, everything is very nice except the bread, butter, tea and coffee... I never saw such places as this for eating, people seem to do very little else.' Breakfast was served at nine, lunch at twelve, dinner at four and tea at 7.30 and the children had their meals at different times. The stewards, she observed, 'seem to be doing nothing all day but laying or clearing tables'. There were twelve stewards at dinner, two for each table.[61]

It is very funny to watch them bring in the dishes: one brings a little bell which the signal for the cook to send up things in the lower regions, then the 12 stand in a row down the saloon passing the dishes like fireman pass buckets; on rough days they have to stand slanting and look very comical, they seldom drop the things; one spilt a cup of tea down a gentleman's back yesterday. Lunch is a pleasant meal of the day, we have hot soup, cold meat, bread and cheese.[62]

She was not impressed by the Sunday service, which was conducted by a 'dreadful old clergyman' who read badly with a nasal twang, a singsong manner and 'has a peculiar habit of raising his voice at the end of each sentence'. The next Sunday there were three services, Church of England, Roman Catholic and a Presbyterian service and 'the captain attended all three so he had enough of it. The quartermaster's came out in a clean white dresses and blue collars and the officers broke out in buttons.'[63]

Captain Gray came in for particular praise from Mrs Compton.

What a very nice man the Captain is, he seems always to be looking out for something to make his passengers more comfortable, he generally chooses these ladies who are travelling alone to walk with among so many because he cannot give much time to each one but he always has a pleasant word or smile for everyone. He has found out a poor little boy of nine years old in the second cabin who does not know a soul on board, his father has married a second wife and is sending him to some friends in England.

Sunday, June 3. The Captain took Joe and me into his cabin one day and into the Cook's galley and the second class saloon. He said he would take me all over the ship when the weather is warmer. He has two cabins, sitting room and bedroom, and everything is so comfortable. The engineer often comes to dinner but none of the mates ever come; Mr Peterson heard of the death of his wife only a few days before we sailed and

I think for that reason he mixes very little with the passengers; he is the first officer, the second is Mr Todhunter, they are both pleasant-looking men.[64]

As they neared Liverpool the bustle began. 'The ship was turned upside down, everybody got into a state of excitement and boxes kept coming up as if there was no end to them, to say nothing of bundles, bags, chairs and portmanteaus.'[65] Finally everyone reached their destined port and thoughts turned from life in the narrow confines of the ship to matters on land.

9

'THIS CELEBRATED STEAM SHIP': FINISHED WITH ENGINES

Before leaving port there was a considerable amount of paperwork and administration to be handled. The ship's register, crew, passenger and stores lists and payment of the various harbour dues had to be taken to the Customs House and then the vessel could be cleared to leave. Add to this the requirements of the emigration authorities and it is clear that much time was taken up in port by administration involving the senior members of crew. The daily Custom's Bill of Entry listed the arrivals of all the ships and their cargoes and these were published in both London and Liverpool. From these the various cargoes of the *Great Britain* can be seen. As only the British lists have survived, the information on cargoes taken out is limited. Items being exported to Australia included clothes, boots and shoes, hats and caps, carpets, haberdashery, vast numbers of slates, earthenware and machinery. Food produce included hams and bacon, beer and wines.[1] But we have little evidence of what was carried out by the *Great Britain* in addition to the luggage of the passengers.

In her early days crossing to New York, she returned with cotton, tobacco, hemp, and various woods such as rosewood and mahogany. On her return voyage in 1853 from Australia she brought goods from Sydney and Melbourne including gold.

Additionally, from the Cape of Good Hope she brought beads, elephant teeth and various other samples. A further stop just before Liverpool was at St Michael's in the Azores and was the only time the ship called there. Here 500 boxes of oranges were loaded for McAndrews, Pilcher and company of Liverpool. The fruit trade was a highly specialised niche market largely dominated by small fast wooden sailing craft that brought the ripening fruit in a race back to the markets in London, Liverpool and Bristol. It may well have been an attempt to test the opportunities, but after 1853 the list of goods arriving on the *Great Britain* were almost all direct from Australia. It largely consisted of wool and frequently cases of 'curiosities' often for museums and specialist collectors. In 1865 she carried 46,88 ounces of gold dust, 524 bales of wool, 693 chests of tea, and 649 passengers.[2]

Safe stowage of cargo was a skilled job. It was important both for the stability of the ship and also any potential damage or contamination of the goods. Some goods had to be kept well apart from one another and there was always the risk of items moving while at sea and packaging of delicate items was crucial. Valuables such as the gold that was regularly shipped had to be kept under lock and key. Some cargoes were inherently dangerous, wool and cotton bales were inflammable and, in the right circumstances, could combust spontaneously. The other problem was that they were bulky cargoes and exceptionally light so had to be packed hard into the hold.[3] Loading and unloading of cargo at Liverpool in the docks was comparatively straightforward but at Melbourne much of the cargo had to be discharged by being taken out of the hold and then over the side of the ship into lighters to be taken ashore which was a slow process.

The various cargoes were destined for named brokers in Liverpool. In 1851 there were around 658 brokers in Liverpool dealing in everything and some specialised in certain commodities such as wool, cotton, corn, leather or in cargo space. There were 11 local banks including a branch of the Bank of England where gold could be deposited.[4] The surgeon, Samuel Archer described leaving Melbourne:

All of course is confusion and bustle by gold and mails coming on board, passengers and their luggage in every direction, friends of passengers etc. We have upwards of half a million of gold and the Royal Mail which should have gone home per S.S. *Simla*. Went to bed in my own surgery rather early. Mr & Mrs Holloway & Miss Southam came on board to wish us goodbye. Before going to bed I cleaned about half the shells which I had found at Geelong.[5]

Captain Gray's replacement as master of the ship was Charles Chapman, aged 52, from Lincolnshire who re-joined the *Great Britain* from the ship *Alsager*. Chapman first went to sea in 1837 in the American trade, then to India and then back to North America. He was first mate of the *United Kingdom* of Belfast from 1852 to 1853, then joined the *Great Britain* as 2nd mate 1853, later becoming first mate, a position he held until January 1869 when he left to become master of *Alsager*.[6]

Sailing of the Steamship *Great Britain*. – Yesterday, about noon, the auxiliary screw steamship *Great Britain*, Commander Charles Chapman, sailed from the Mersey for Melbourne, with a full complement of passengers and very valuable general cargo. As the vessel gently steamed down the river her fine proportions and frigate-like appearance were greatly admired by a crowd of spectators who witnessed her departure from the pierhead. A party of gentlemen connected with the trade of the port accompanied the steamer down channel, and subsequently returned on board the tender. The weather was brilliant, with a gentle breeze from the west.[7]

The *Great Britain* as an auxiliary steamer had done great service for Gibbs, Bright, but technology had now advanced so that full steam power could be considered. The biggest challenge for the Australia run had been in achieving it in constant steaming. This was an aim that Brunel and many others had failed to achieve. The problem was that the engines needed so much coal that regular re-coaling visits had to be made en route. This is why the sailing ship with auxiliary steam had become the preferred type of vessel, being able to take the benefit of favourable winds without using coal and to use steam when there was no wind. While much attention had been paid to the structure of the vessels and the nature of propulsion, the way in which the engines consumed coal was now a focus. Francis Humphreys and younger brother Edward Humphreys were engine builders. Francis, as seen previously, had worked on the early engines for the *Great Britain*. His younger brother was now looking at experimenting with two-cylinder engines for P&O. In 1861 these double cylinder engines were more economical with fuel, but were accident-prone. Randolph Elder took out a patent in 1853 for a double-cylinder marine engine. It was Alfred Holt who installed compound engines in 1865 for a new line of cargo steamers from Liverpool to Singapore. By the 1870s the engine improvements meant that constant steam to Australia was achievable.[8]

Gibbs, Bright now developed a new venture together with a man very well known to Brunel, Daniel Gooch, and with the idea first mooted by Brunel in 1836 of combining rail and sea travel. In April 1873 the Anglo-Australian Steam Navigation Company Limited was announced with a proposed capital of £1,250,000 in £10 shares. The directors' names were headed by Sir Daniel Gooch, MP and Chairman of Great Western Railway Company. Gooch had known Isambard Brunel for some years, having been appointed by him to the Great Western Railway. Charles Lamport was a shipbuilder and Tyndall Bright was the director representing Gibbs, Bright & co. The directors included four other Members of Parliament: Lewis Llewellyn Dilwyn, who was also a director of the Great Western Railway, Charles Gilpin, who was also Chairman of the National Provident Institution, and Thomas Brassey MP, son of the railway pioneer.[9]

The managers of the new line were Gibbs, Bright and Co. Samuel Bright had died in January 1870 and his two eldest sons, Henry and Heywood, were in the Liverpool firm while his youngest son, Samuel, was a partner in the firm of Bright Brothers and Co. in Melbourne.[10] Anticipating considerable trade, the new company also had an impressive list of brokers: Mullens, Marshall and Co., J. and A Scrimgeour of London; G. and T. Irvine, Robert J. Tinley of Liverpool; Warner and Page of Manchester; Kerr, Anderson of Glasgow; Bruce and Byrnes of Dublin. The secretary to the company was James Kenworthy and their offices were at 4 Lothbury, London and at 1 North John Street. Liverpool.[11]

The prospectus laid out their plans to establish a line of full-powered steamships between Great Britain and Australia, via the Cape of Good Hope, for the conveyance of 'Passengers, Mails, General Merchandise, and specie'. They planned a monthly line each way to be served by six vessels of 4,000 tons and 'fitted with the latest machinery'. They confidently predicted the passage to Melbourne would be achieved in 42 days without stopping. Each ship would carry 100 first, 100 second, 106 third Intermediate, and 300 steerage passengers.[12]

Links were made with the colonial steam-ships for direct communication with the various ports throughout Australia and New Zealand, many of which were owned by Bright Brothers of Melbourne. They pointed out the amount of trade between Australia and Great Britain. The 'enormous difference in emigration between the American and Australian Colonies', the prospectus declared, was due to the lack of regular steam communication between Britain and Australia and quoted the 'Report of Sir George Verdon, the late Agent-General of Victoria'.[13]

When once on board ship, perhaps, the length of the voyage is not of great importance. It is the certainty of the time of starting, the knowledge that it is being made as quickly as possible, and that it will end in so many weeks, that gives a steamer so great an advantage over a sailing ship.[14]

The authorities in Victoria 'have intimated they are prepared to subsidise the Company on its formation, making a contract with them for a Mail Service for a period of five or ten years'. This was, as ever, the goal of many shipping companies, to win an operating subsidy.[15] In Britain the departure ports selected were Liverpool and Milford, both being well served by railways and by steamships with Ireland. The Great Western Railway Company would be prepared 'when the steamers are ready' to run special trains from London to Milford in eight hours.[16] 'Gibbs, Bright and Co., of Liverpool and Melbourne', with their considerable experience in the working of 'Auxiliary Steam' to Australia, would manage the ships. Additionally, it was announced; in order to maintain 'for the benefit of the Company, the existing connections for freight and passengers until the new ships come on the berth' the *Great Britain* was to be purchased by the Company, 'at a valuation, the purchase money being represented by fully paid up shares'.[17]

Meanwhile the *Great Britain* was heading for a new destination. For the vast majority of voyages the *Great Britain* carried passengers and cargo to and from South Australia. In 1859 Queensland became a separate colony with large amounts of uninhabited land and the need for immigrants to help develop its economy. From 1861 they came up with a system of land purchase to finance emigration. Led by a man called Henry Jordan, there was a concerted effort to promote the new colony and to encourage migrants to choose Queensland rather than other parts of Australia. Jordan persuaded Liverpool ship operators to lay on sailings and to sell passages to Queensland.[18] So the Liverpool and Australian Navigation Company advertised on Saturday 25 October 1873 the 'celebrated auxiliary steamship' under the command of Charles Chapman to sail for Melbourne and Brisbane. It would also take passengers for Sydney, Adelaide and New Zealand, although fares were only advertised for Melbourne and Brisbane, as passage to the additional locations would be met by local steamers, of which Bright Brothers in Melbourne had several.[19] But this additional port of call did not bring a new flood of passengers or cargo.

The ship continued to sail from Liverpool until June 1875 when she was advertised as leaving from London. 'The Liverpool and Australian Navigation Company's celebrated steam ship' was announced to be leaving from South West India docks on 23 August and also picking up passengers from Gravesend. She was heading for the principal ports in Australia and New Zealand.[20] However, the ship sailed without Captain Chapman, who never reached London as he died at his home in Toxteth of dropsy on 21 June 1875. 'Captain of the *Great Britain*. Deeply regretted, Melbourne papers please copy.'[21] First officer Peter Robertson again had to step in and this time he remained as master of the ship, but her Australian days were numbered.

Having been launched with the usual fanfare to entice investors, The Anglo-Australian Steam Navigation Company Limited did not feature again in the news and did not appear to take over the *Great Britain*. Competition for direct steam to Australia came from other directions. Yet another company announced its plans, this time already with a ship and with Gibbs, Bright as its agent.

THE AUSTRALIA DIRECT STEAM NAVIGATION COMPANY (Limited). STEAM to AUSTRALIA under 45 days, leaving London on 20th March, and Falmouth on 34th March, for MELBOURNE direct, forwarding passengers and goods at through rates all parts of the Colonies, Tasmania, and New Zealand. – The Australia Direct Steam Navigation Company (Limited) will despatch their splendid full-powered Screw Steamship VICTORIA, 22 years A1, 3,985 tons register, and, 2,000-hp. effective (the pioneer of a regular monthly line), to load in the South West India Dock, and will call at Falmouth to embark passengers. An experienced surgeon and stewardess will be carried. Fares: First-class 65 guineas, second-class 30 guineas, third or intermediate 18 guineas, and steerage 12 guineas. Return Tickets (first- and second-class only) will be issued fare and a half, and a fare and two-thirds, as may be arranged. For plans of cabins, and rates of passage money, apply to the various Agents throughout the country and abroad; in Liverpool to GIBBS, BRIGHT & CO.; London, at the Offices of the Company, New Broad Street, E.C.; and for freight and passage to the Broker, JOHN BENNETT, June., 1 East India Avenue, Leadenhall Street, E. C. The VICTORIA will leave Melbourne on her return voyage early in June. Agents in Melbourne, BRIGHT BROTHERS & CO.[22]

The ship was previously named the *Nebraska* and had been built in 1857 in Jarrow. The company was chaired by Admiral Frederic Crauford and had only recently been established. It proposed to raise £500,000 but never did so. The ship was purchased and brought to London but never left. Tickets had been sold, but the company went into liquidation. Several court cases followed from aggrieved potential passengers and the master who claimed for his wages. There was one case against the well-established and well-respected broker, John Bennett, who had sold tickets on behalf of the company.[23] It was not good publicity for Gibbs, Bright, who had previously survived the crash of the Australian and Eastern Navigation company, which had been planned as a merger of the Black Ball, White Star, and Gibbs, Bright's Eagle Line. That affair in 1865 had become a financial scandal on the Liverpool Stock exchange with blame largely landing on McKay, owner of the Black Ball Line.[24] Gibbs, Bright's troubles had not ended there. In February 1868 they went to court to obtain £18,750 from the liquidators of Barned's bank and won their case, but their main rival, the White Star line, was brought down by the crash of Royal Bank of Liverpool. With emigrant numbers continuing to fall it was not a good time to be in the Australian business.[25] Gibbs Bright managed to continue their Liverpool to Australia service with the *Great Britain*, but 33 years after her launch, the *Great Britain* returned to Liverpool in February 1876 on her very last passage from Australia.

For five years the *Great Britain* lay idle at Birkenhead until in July 1881 she was put up for auction.[26] She was advertised by the auctioneers Messrs Kellock & co as a ship with a 'history of more than ordinary interest... Well-known by her passages as a most successful ship'. The sale attracted a 'very large attendance of gentlemen who are closely identified with the shipping interests in the port'.[27] It was suggested she could be suitable for the cattle trade and carry livestock on three decks. At the auction, while attracting considerable attention, she did not achieve her reserve price as the highest bid was just £6,000. But as before, she could still attract rumours.

A Floating Hotel for Southend

A company is about to be started for the purpose of establishing a Floating Hotel and Sanitorium at Southend. The prospectus which has been issued states that the

"Great Britain" steamship has already been purchased with the object of fitting her up for the purpose, and mooring her off the Pierhead. It is proposed to provide accommodation for a thousand persons, with the convenience usually obtained at a first class Hotel on land. The ship is to have the electric light at the mast head at night, and thus render service as a beacon. A steam launch is also to be in constant readiness for the convenience of those who prefer a "life on the ocean wave" and other inducements too numerous to mention. We may reasonably ask, what next?[28]

Indeed, it was a good question, but it was yet another scheme that came to nothing, as the ship had not been sold. There was another auction on 20 February 1882 to sell the considerable quantity of stores, china and furniture from the ship. Cutlery, glass and china, cooking utensils, mahogany dining tables, settees, 80 mirrors, bookcases and 500 books, sideboards and armchairs, all went under the hammer. Table and bed linen, carpets and curtains were all for sale. The souvenir hunters could purchase soup tureens, entree dishes or tea and coffee pots, or even one of the many pieces of electroplated silver, from fish carvers to napkin rings and toast racks.[29]

By now the two branches Gibbs, Bright & Co in Liverpool and Bristol and Bright Brothers & Co in Melbourne, Sydney and Brisbane had separated and it was announced that the Australian business would continue led by Tyndall Bright, Charles Edward Bright, and Reginald Bright. In 1881 Gibbs, Bright and Company including its remaining plantations in the Caribbean was absorbed into the London based Anthony Gibbs & Co.[30] Anthony Gibbs now acquired the redundant *Great Britain*. Anthony Gibbs mainly traded with North and South America. Unable to sell the ship for a suitable price, the decision was taken to use the good-sized vessel as a cargo ship in their wheat trade.

She was rebuilt by Grayson's, the long-established shipbuilders in Liverpool. The engines were removed, which expanded the cargo space and all passenger accommodation was taken off, as were the deck houses and superstructure. Her hull was sheathed in pitch pine and then covered again in turn with zinc, the latter mainly to protect it from worm. The aperture for the propeller was plated in to improve the ship's steering. The ship was registered again in Liverpool on 10 November 1882 as an iron sailing ship and she now had a Lloyd's classification and was given the class of A1. She was now a sailing cargo ship and her new master was named as James Morris.[31]

James Morris was a Danish mariner age 38. He was born in Kirkholm, Denmark, but gained his mate's certificate and master's certificate in Liverpool in 1868 and 1874 and was now based in Liverpool. He worked on Danish vessels from 1862 until 1864 then British-owned ships. He had a short stay as master of the *Great Britain*. He recruited a crew but the ship remained without a cargo, so Morris left to join another ship.[32]

The next master was F Kerr who was employed as master for one day before his place was taken by Henry Stap with Kerr now as first mate. Henry Stap officially took charge on 2 November 1882. He was somewhat older than James Morris at age 55 and considerably more experienced in long distance voyages, having taken the route from Liverpool to Australia for many years. He was a Yorkshireman originally from Skipsea but now living in Wallasey. His time at sea began as an apprentice in 1845 in London-registered ships and he served as a mate from 1850 to 1852. He was the second mate on the *Mariner* of London in March 1852, but a handwritten addition to his career notes adds 'reported as deserted in Port Philip in 1852'. It looks as if he was tempted, as so many seamen were, by the opportunities in Australia. The new opportunities did not work out and he went back to sea working on Australian coastal vessels a few years later. By 1855 he had returned to British vessels and worked his way from able seaman to mate again by 1857. He was married in 1861 age 32 and had three children, two of whom were born at sea.[33]

The *Great Britain* loaded a cargo of 3,290 tons of coal and headed off in late November aiming for San Francisco but was forced to turn back and returned on 24 November owing to leakage. After repairs and another survey, the ship set sail again on 2 December. She eventually sailed for San Francisco in December but had to put into Montevideo becuase of problems with the stowing of the cargo, which had caused the ship to be unstable. The crew apparently 'refused to proceed until the vessel was surveyed and lightened'. She spent 30 days in Montevideo where 200 tons of cargo were removed and 500 tons shifted to a lower deck. The ship arrived finally at her destination on 2 June 1883 after a 180-day journey from Britain. She was, nevertheless, still held in high regard, the local paper, the *San Francisco Alta*, referring to her history:

> The British ship *Great Britain* docks today at Mission Street wharf to discharge her cargo. This famous old ship is well worth a visit and she carried us back to auld lang syne and shows conclusively that they put good work and good materials in vessels in early times. It is related of this vessel that during the early part of her career she was sunk in Dundrum Bay and was floated off by the normal process of filling her with bulrushes. Captain Stap, her commander, is one of the best ship masters in the service and we hope his stay among us will be a pleasant one.[34]

Many of the crew clearly hoped for an extended 'pleasant' stay and deserted, in both Montevideo and San Francisco. With a largely new crew, she sailed for home one month later carrying wheat for Cork and again it was a long journey, this time taking 154 days. She returned without the first mate, Kerr, who was discharged in San Francisco by Stap, 'by mutual consent due to drunkenness and neglect of duty'.[35] The cargo was eventually discharged at Liverpool.

She departed again in May 1884 with 2,870 tons of coal for San Francisco and encountered bad weather. They met a south-westerly gale near the Falkland Islands and after struggling around Cape Horn arrived after a voyage of 160 days. These passages around the unhospitable Cape Horn were exceptionally hard on ships and men as they faced the severe weather conditions in the South Atlantic. She returned from San Francisco on 12 February 1885 carrying 3,440 tons of wheat and arrived in Cork 145 days later, before docking at Liverpool on 12 July. Here she remained until the next year when she sailed to Cardiff.[36]

The ship could still turn heads and attract attention even in such an incredibly busy port. In the *Western Daily Press* on Thursday, 21 January 1886 three brief sentences are squeezed in between an obituary for a local worthy and a review of some amateur dramatics at Clifton. 'The *Great Britain* steamship, built at Bristol many years ago, entered Penarth Dock yesterday. After being a passenger boat for many years she is now to receive a cargo of coals. She is still a trim-looking vessel.'[37]

The *Great Britain* left Penarth on 6 February 1886 and a few weeks later on 25 March met heavy weather in the South Atlantic. It was an eventful journey. A small fire was discovered, which sounds rather as if a crew member had been smoking below, and this damaged the ship's stores and burnt some sacks and ropes before it was put out. The ship proceeded on towards Cape Horn where on 16 April they met a south-westerly gale that strengthened to hurricane force. The ship began to leak badly and the crew were alarmed. Concerned that the ship was straining too much, they protested to the captain who still decided to continue on. But the cargo of coal had shifted causing the ship to list to port, so the long-suffering crew had to shovel coal to redistribute its weight evenly and the ship was eventually levelled. On 3 May another gale struck the ship and this time lasted a week and by 11 May it had again become a full hurricane. When the crew again protested Captain Stap finally agreed to turn back and headed to Port Stanley in the Falkland Islands, arriving in the afternoon of 24 May. She was labouring and touched ground twice and eventually came to anchor in Port Stanley.[38]

A special survey reported that the hull was 'quite tight', but the damage to the spars, rigging and deck were estimated at £5,500. The ship was now at the other end of the world in a place where repair facilities were limited. The aging ship had not so far proved a great success as a pure cargo ship and it was not worth the company's money or efforts to repair her and return her to England.

With this situation now the Falkland Islands Company began a purchase negotiation with Anthony Gibbs & Company. Gibbs estimated the assets and value of the ship (excluding the cargo) at £3,000, a figure which the Falkland Islands Company initially deemed excessive, but eventually they agreed following a meeting of their Board of Directors.

> Dear Sirs, I have the satisfaction of informing you that our Directors have instructed me to hand you a cheque for £3,000 tomorrow in exchange for a Bill of Sale for the *Great Britain* and a letter in duplicate to Captain Stap telling him to hand over the vessel, her furniture and all stores and effects as in existence on the 19th November, the date of purchase of the vessel.[39]

With the sale the crew could now be released from their contract and discharged and try to get back to their homes via another ship. This could take some time, depending on which ship called at Stanley. Even after the sale agreement, debates continued between the two companies for another week, to agree the duty paid for holding the coal cargo in Stanley and the balance of provisions included in the sale. Finally, the Falkland Islands Company agreed to purchase the coal for an additional £1,750.[40] On 8 November 1886 Anthony Gibbs sold the *Great Britain* to the Falkland Islands company and her registry was closed on 19 July 1887 when she was converted to a hulk. With the coals removed, the *Great Britain* now became a floating wool storehouse in Port Stanley, wool being the mainstay of the Falkland Islands economy.

This arrival in the Falklands was not the ship's first visit. She had arrived in Port Stanley during the course of a run to Melbourne on the return leg of her 10th voyage under the command of Captain Matthews. During her four days there over the New Year in 1853/1854 one of her passengers did some exploring and was unimpressed, subsequently writing an unflattering account of his visit in an anonymous article in the *Bristol Mirror* on 25 February 1854:

> At daylight in the morning of the last day of the year 1853 we sighted the beacon at the entrance of Stanley Harbour... After proceeding about three miles, we took a pilot on board, and passed through a narrow entrance less than 200 yards in width, and at 10am, the '*Great Britain*' dropped her anchor

opposite the township. The distant country looked dreary enough, and reminded me very much of the rolling hills of Morocco. The total absence of trees or shrubs gave it a wild and desolate appearance... I wish I could describe to you my sensations of the dreary and wild nature of the country, as I walked with my gun about the ruins of the houses and fort at Berkeley Sound. It seemed to me as if the place was under the spell of some potent magician, so desolate was the scene... The inhabitants were very anxious that we should give a favourable account of their island; but I must confess that I cannot anticipate a glorious future for the Falkland Islands, or that Stanley Harbour will ever be anything more than a coaling depot for steamers, and a harbour refuge for disabled vessels. When the 'Panama Route' is opened, it will be to them a heavy blow, and great discouragement. After three days detention, during which we took in 300 tons of coals, we bade adieu to the Falkland Islands without regret.[41]

Meanwhile, in 1886, as the *Great Britain* was sold to the Falkland Islands Company, the Anglo Australian Steam Navigation Company that was so heavily advertised in 1873 at last had two passenger ships built for it by Messrs Leslie and co on the Tyne, but it was still a tough trade. Despite the new full-powered steel steamers that were 'confidently expected to make the passage to Adelaide within 43 days' fares from Britain were reduced.[42]

In the same year it was announced that the following partnership was dissolved. Henry Hucks Gibbs, Augustus Sillam Alban, George Henry Gibbs, James Charles Hayne, Vickery Gibbs, Herbert Cohayne Gibbs, Tyndall Bright, Charles Edward Bright and Reginald Bright trading as Anthony Gibbs, Sons and co, Liverpool and Bristol, and as Gibbs, Bright and co in Australia and New Zealand. It was the end of a long era for the Gibbs and Bright families.[43] Tyndall Bright died in 1902 in Bristol and Charles Edward Bright, founder of Bright Brothers in Melbourne, who had travelled so often on the *Great Britain*, died in 1915 having retired to England.[44]

The ship now was a storage hulk for the main commodity of the islands, wool, but even this peaceful and static occupation was threatened when there was a dispute between the Falkland Islands Company and the colonial authorities. The colonial administration proposed the building of a pier to make loading and unloading wool easier. This would effectively render the *Great Britain* redundant. The Falkland Islands Company saw this move as a threat to their storage services for shipping companies.

> We have shipped on more than one occasion over 2,000 bales by one steamer, and for such a quantity the *Great Britain* has ample capacity; it is evident, therefore that the rival Government establishment must provide equal accommodation. Bales of wool and skins cannot lie for weeks exposed to the weather in one of the most uncertain climates in the world and storage for 2,000 bales at least must be provided.[45]

The idea of a proposed pier was abandoned and the *Great Britain* was safe to continue her storage role sitting in the harbour at Port Stanley.

10

'A UNIQUE NAUTICAL ANTIQUE': IN THE TWENTIETH CENTURY AND TODAY

By the end of the 19th century, all of Brunel's big ships had long since ceased operating. The *Great Western* was broken up in 1857 and the *Great Eastern* in 1888. The *Adelaide* and *Victoria* were both sold in 1860, the former was noted as being 'sold to foreigners', and newly registered as the *Mersey* in Liverpool.[1] The fate of the *Victoria* is unknown. The *Great Britain* remained by an accident of fate, abandoned and forgotten in a remote part of the South Atlantic.

Even in the early years of the twentieth century she might occasionally be remembered with nostalgia. Field-Marshal Sir Evelyn Wood wrote to an old friend, Mr James Steers of Bath, to recall their time on board the *Great Britain* when they sailed from Ireland to India. He remembered the ship was 'ship-rigged, and steamed well sailed with canvas... On this voyage she made Bombay in 70 days from Queenstown'. The fate of Captain Gray was still a talking point: 'Your letter brings back me the time when we sailed together on the "Great Britain." I have for the moment forgotten the name of the skipper – now I have it – it was Gray; he came to an untimely end and a watery grave, for he threw himself into the sea. It was the effect of the heat, I believe.'[2]

The ship remained afloat in Stanley harbour in the Falkland Islands in the ownership of the Falkland Islands Company. During the First World War officers from the British fleet were said to have visited the *Great Britain* and she was something of a tourist attraction for any visitors.[3] But by the 1930s, even her days as a floating warehouse were over. In 1933 the ship's weather deck was leaking badly and she could no longer be used to store wool and hides and a debate now began over her future. People had very different ideas. The Port Stanley harbourmaster saw the ship as a dangerous obstruction, his primary duty was the safety and protection of the vessels using the port. In February 1936, his proposal was that the *Great Britain* should be destroyed. In his view 'the proper course would be to tow this hulk out to sea and sink it at a place where the waters are of such a depth the sunken hulk would not become a danger to shipping.' And he was keen to see it sunk well outside the main shipping routes to the island. There were even suggestions that she might become a useful target for the Royal Navy to practise gunnery.[4]

The Governor of the Falkland Islands had a different view. Sir Herbert Henniker Heaton was appointed as Governor and Commander-in-Chief in January 1935. A very experienced diplomat, he was no newcomer to the Islands having previously served as Colonial Secretary in 1921 and subsequently acting governor in 1923 and 1924. Heaton saw the significance of the ship to British history and was keen to save her for posterity, but it was not a widely accepted view locally.

> ...it was Heaton, rather than involved community participation, which drove the thwarted restoration attempt ... it is also apparent that this proposal largely ran counter to the wider interests of the Colonial Administration, and the more prosaic concerns of many Falkland Islanders themselves. Despite being relatively short-lived, Heaton's proposal prompted a considerable degree of opposition, which presented strong arguments against the ship's continued presence in Stanley which was deemed detrimental to shipping interests and the wider Falklands economy.[5]

While the harbour master at Stanley insisted the vessel should be destroyed, seeing it purely as a wreck and a danger to commercial shipping, Heaton suggested that some attempts should be made to preserve the vessel and estimated a cost of £3,000. He wrote on 3 March to the Governor of the Bank of New South Wales in London, planning to open up an account linked to a public appeal.

> Dear Sir
> I propose to make a public appeal through the press for funds for the preservation and restoration of the ship *Great Britain* now lying as a hulk in Stanley harbour.
> As you are doubtless aware this vessel was the biggest ship in the world when she was launched at Bristol in 1845. For a number of years between 1852 in 1875 she was the most celebrated steamship on the Australian run.

He proposed that the rig be restored to the one she wore when she was running to Australia and finished by pointing out that 'she is by far the longest-lived iron ship in the world'.[6] He also wrote to the Lord Mayor of Bristol urging his support, but had no useful response. Locally, he asked the Falklands Director of Public Works to produce a feasibility study for the restoration. The resulting 27-page document outlined the labour and materials required for such a restoration and the specialist and expert requirements. On 12 April 1936 the minutes of the Falklands Executive Council show the result of their deliberations:

> With reference to the disposal of the *Great Britain* His Excellency explained that the Falkland Islands Company Limited made an application for permission to beach the vessel in Sparrow Cove in the outer harbour. He had intended appealing for funds in order to preserve her. The Director of Public Works who had made a thorough examination of the hulk had reported it would cost £10,000 to place it in a state of preservation. His Excellency was therefore forced to abandon the project with regret.[7]

Heaton had been unable to find anyone prepared to take the ship, ideally back to Britain, even for free, and the harbourmaster was still keen to destroy the ship. Heaton could not take such a permanent step and he ordered the ship to be taken to Sparrow Cove and in doing this managed to save the vessel.[8]

The lack of response to Heaton's brief appeal to bodies in Britain came at a time when HMS *Victory* was finally being preserved in Portsmouth dockyard and the *Cutty Sark*, with all its romantic tea race connections, was being kept afloat by one dedicated owner. Both of these were sailing vessels and there seemed to be less interest in a rusting relic of Britain's industrial past, which was sitting many thousands of miles away.[9]

The *Great Britain* was towed to Sparrow Cove, 3½ miles from Stanley, where she was run aground and scuttled. There she remained, just one of the many ships abandoned in the islands.[10] As one Falkland Islander later pointed out, 'Have you ever seen an early photograph of Stanley Harbour ... it was dotted with them [ships] all the way down to small boats, y'know. Dozens and dozens of them.'[11]

During the Second World War she was still an object of attraction for those servicemen passing through the Falklands.[12] But it was the 1950s that brought a reawakening interest in the *Great Britain* and it came from two quite different sources. One man saw the international significance of the ship. Karl Kortum was the Director of the San Francisco Maritime Museum and he wrote to Frank Carr, the Director of the National Maritime Museum at Greenwich, and suggested that there should be some kind of effort to rescue the vessel. There was no positive response and Kortum began to look for backing to bring the ship to San Francisco. Meanwhile in Britain in 1952 Ewan Corlett, a naval architect, acquired a copy of the painting by Joseph Walter showing the *Great Britain* in 1845 during her trials in the Bristol Channel. It was the start of a lifelong obsession with the ship.[13]

Seven thousand miles away the ship's move to the calm waters of Sparrow Cove did not mean that her useful life was completely over, as she became a source of seafood. Locals spent many happy hours foraging for the mussels that populated her hull. As one Falkland Islander vividly recalled:

> When we used to get the mussels we had a long pole about 10–12 feet long and it had on the end of it, it had kind of a garden rake and under it a bit of wire mesh ... like a sack and you put down over the side of your boat until you've got down below the mussels and you kind of scratched your way up the ship from the bottom to the top ... and then you lifted it up and see the mussels and sea urchins and little crabs and god knows what. So you sorted out the good mussels and put them back in the water. It would take you four hours to get a couple of sackfuls, big sackfuls and all those big blue mussels. And now that's she's gone you can't have them.[14]

It took another decade before any progress was made to save the ship and it came when there was growing concern to preserve the industrial archaeology of Britain, during a time when modern brutalist concrete structures were being built and old Victorian structures were being swept away.[15] Corlett wrote to *The Times* newspaper in November 1967, urging action to save the ship for the nation and situating it alongside the *Cutty Sark* and HMS *Victory* as national treasures. He wanted to gather support to retrieve the ship from the Falklands and return it to Britain. Realising the sheer scale of such a task, he suggested that the ship should, at least, be fully documented and photographed if it could not be preserved. Before his letter to *The Times*, Ewan Corlett had done his best to assess the state of the ship even though he had been unable to get to the Falkland Islands; certainly he knew that an iron ship in saltwater might not last very much longer.

Corlett was unaware at the time of the interest from the San Francisco Maritime Museum.[16] There was yet a third expression of interest in the ship. Peter Lamb was a model maker from Stratford-upon-Avon and he wrote to the Colonial Secretary.

> The *Great Britain* seems to be in an amazing state of preservation and, on reflection, it seems a pity that the

appeal to restore her in 1937 was not made at the present time, when people are far more preservation minded. I wonder indeed if such a fund was indeed started today, whether the hull is capable of being preserved at this late stage?[17]

The seemingly innocent query letter to the Colonial Secretary was actually linked to interest from Robert Vogel, the curator of the Smithsonian Museum in Washington.[18]

Corlett's letter to *The Times* triggered a very positive response and among them was one from the Falkland Islands Company, who had been the previous owner of the vessel, and also from the National Maritime Museum. While Karl Kortum's letter to the Greenwich-based museum some years before had not led to action, the current director of the museum was Basil Greenhill. Basil had served in the diplomatic service and, although not by background a historian, he had an instinctive feeling for old ships and in 1952 had written one of the great works on merchant sailing ships.[19]

An important meeting was held at the office in London of the Falkland Islands company on 5 April 1968. Ewan Corlett, Basil Greenhill and two directors of the company were joined by Richard Goold-Adams. Richard was to become another key player in the whole rescue project. Corlett's letter to *The Times* had prompted an article about the ship in the *Observer* and it had included a striking photograph of the ship as she lay in Sparrow Cove. It was this that had spiked Goold-Adams' curiosity. A strong patriot, he felt a particular attraction to a ship that bore the name *Great Britain*. By background he was a journalist, having written for the *Economist*, and he had also been involved in television news at ITN. He co-founded the International Institute for Strategic Studies, which exists today to provide independent views on world security. He was a mover, shaker and influencer with a wide network of very well-placed contacts.[20]

The meeting was given an assessment by Corlett of the condition of the ship. She was in danger of breaking up because of a major crack in the starboard side. Due to her situation in

Sparrow Cove and the tidal flow, the bow and the stern of the ship were unsupported, which put an enormous strain on the centre of the ship. This was its major weakness, but Corlett was still of the belief, without seeing the ship, that she could be brought home, even though this meant 7,000 miles of towing across the Atlantic. Despite the relatively gloomy picture and so many unanswered questions it was agreed that further steps should be taken. At this stage they became aware of the interest in the ship from the Director of the San Francisco Maritime Museum, Karl Kortum.

The next meeting one month later was held in Bristol under the auspices of the newly formed Brunel Society, a local Bristol society formed to celebrate Brunel's engineering achievements. The main financial backer for the San Francisco bid, Bill Swigert, flew from San Francisco to attend.[21] Goold-Adams described the meeting and the concerns at this early stage:

> There was some anxiety about the intentions of the Americans. Bill Swigert was very open and happily shared his considerable knowledge with the meeting but still the question remained. Finally Bill provided clarity. 'I think I can speak for all of us in San Francisco, when I say that we feel that the right place for the *Great Britain* is here in England where she was built.' And he paused for effect 'If a serious British effort is going to be made to salvage the ship, therefore, we in America will stand back and wait for you to go ahead.' And he concluded by putting all of the considerable research, much of which he had sponsored, at the disposal of the new committee. It was an extremely generous gesture. They had been looking at this for some years and had spent considerable time, money and effort and were now stepping back to allow a very new, untried and untested team to take the lead.[22]

The embryo *Great Britain* project was still just an assembly of a few enthusiasts, a steering committee chaired by Goold-Adams. Basil Greenhill suggested that the Society for Nautical Research become the initial host body for the project and provide the project with a treasurer. The society was established in 1910 and had played a major role in the rescue of HMS *Victory* and was happy to help. Basil, himself, always remained a committed supporter of the project and would later take over from Goold-Adams as the chairman of the project when he retired from the National Maritime Museum.

Meanwhile the committee was moving at speed and by September had organised a press conference to launch the project. At this stage Ewan Corlett's estimate of the salvage cost was £150,000, so fund raising was critical.[23] Goold-Adams was highly effective at pulling together the right people. With his knowledge of journalism, he had early contact with the BBC who were interested in producing a documentary about the project. The project also had good contacts within the Labour government of the day. These were to prove vital. Denis Healey who was then the Minister of Defence and the Under Secretary to the Navy, David Owen, were sympathetic and with their tacit support, the Royal Navy, as it had in the previous century, was to be of considerable help. A survey had been carried out on the ship by a local detachment of Royal Marines at the request of Sir Cosmo Haskard, the Governor of the Falkland Islands. A series of close-up photographs were also especially commissioned.[24]

The crucial matter was to get their technical expert, Ewan Corlett, to see the *Great Britain* for himself. Transport to the Falkland Islands was not straightforward and Corlett was a busy man who could not afford to leave his consultancy practice for any length of time. With sympathetic assistance, the government, through Denis Healey and David Owen, were able to exploit an opportunity. Lord Chalfont, who was Minister of State at the Foreign and Commonwealth Office, was due to make a visit to the islands on HMS *Endurance*, an ice patrol ship. Corlett flew to Montevideo and embarked on HMS *Endurance*. It took four days to get to the islands and during that time Corlett won the whole ship's company over to the cause and, as a result, had many men keen to volunteer to give unofficial help.[25] So Corlett aided by the ever helpful Royal Navy was able to take detailed measurements of the ship and assess the situation.

Every morning at 8.30 the survey party of Corlett and his naval volunteers left the ship and were escorted by porpoises, seals and penguins to Sparrow Cove.[26] In addition, lights, ladders, hammers and diving gear were loaned by HMS *Endurance*. The BBC was on hand with a camera team to take some preliminary footage. Corlett also took the opportunity to meet with the governor of the island and from that meeting came away with the impression that a plan to move the ship was not going to be very popular with the residents, who saw it as their treasured possession. Despite this Corlett, ever the practical man, at the end of his stay he returned to write a detailed report on the condition of the ship and the possibility of re-floating her. The really good news was that the hull was in considerably better shape than the doom mongers had predicted. It was an old theme replayed – the *Great Britain* continued to surprise her detractors.[27]

The scheme now looked possible and they even had a potential home for her, her original dock in Bristol. Richard Hill, chairman of Charles Hill & Sons Ltd, held the lease of the Great Western dockyard, which was the original building dock for the ship. He was very keen that this should be the permanent home of the salvaged vessel.[28]

The project team, in its rush of enthusiasm, had overlooked the need to approach Bristol City Council at an early stage. The council owned the lease. Containerisation was now focussing shipping at Avonmouth and the Bristol docks were run down and out-of-date. The Council wanted to redevelop the port area and had no plans to have a rusting ship stuck in the middle of their bold new urban environment.[29] Once there it might never get out again, they could not see a financial future for the ship and feared it would be at risk to Bristol ratepayers.

The project contacted Bristol Corporation in November 1968 but the Council did not see it as urgent. Their view changed in March 1970 when it was made plain the salvage was actually underway.[30] With some persuasion it was finally agreed that she would return initially to Bristol.[31] At this stage, two more people joined the steering committee, Maldwyn Drummond, a keen yachtsman and a founder of the Maritime Trust, and Lord Strathcona, a Conservative peer and steam enthusiast.[32] Both would continue to be energetic and influential supporters.

A key question was would the ship, after so many years and several alterations, fit back into her original dry dock? She was now somewhat wider having had additional timber fitted over the iron in her last refit. A template was created to check the size of the dock and the team confidently predicted that the *Great Britain* 'would *just* fit in'.[33] The organisation now needed to raise the considerable funds to achieve its goal and, in order to do that, sought professional help. Up till now it had been a group of busy influential men with their own careers, but the volume of correspondence and business was growing and simply could not be handled just by volunteers. The first general secretary appointed was Dermot Fitzgerald Lombard, a retired Royal Air Force Wing Commander, and offices were provided just off Chancery Lane in London.[34]

Many things were coming together, including a Royal Mail stamp issue of great British ships which portrayed the *Great Britain* on the one-shilling stamp. All of this was helping to increase the public awareness of the *Great Britain*. They also managed to engage one very well-placed and critically important supporter. HRH, The Duke of Edinburgh sent a message of encouragement:

> The transition from sailing ships to power driven ships was the most profound in thousands of years of maritime history. Nothing quite like that revolution will ever be seen again. The *Great Britain* represents a vital stage in that revolution and is therefore of immense interest to future generations. I very much hope that the attempt to bring her home and put her on display will be successful.[35]

The *Great Britain* had been registered under several owners during her service life. But a single strand linked most of them, the Gibbs family. The last official registered family owner in the 1880s was Vicary Gibbs. In 1969 there was still a

Mr Anthony Gibbs in London in the bank that bore his name, and he was naturally approached for help.[36] But not everyone was enthusiastic about the project. An automatic assumption was that money could be raised through the shipping industry, but that was swiftly disabused. The shipping industry, in the midst of its own containerisation revolution, was not interested in looking backwards and being connected with old ships.[37] Neither was there a warm feeling about matters in Bristol, as one Bristolian pointed out to Goold-Adams.

> I'm afraid you won't find Bristol people falling over themselves to help you. You see most have never heard of the *Great Britain*, and those who have don't take it seriously... I'm afraid you got off to a bad start here, and that's the fact of it.[38]

David Owen MP had been of assistance in getting Ewan Corlett out to the Falklands and he was responsible subsequently for the introduction to the project of Jack Hayward. Hayward was an engineer and philanthropist and passionate about British heritage, having given £150,000 to buy Lundy Island for the National Trust. Hayward's father, Charles, began his career as an apprentice patternmaker in Wolverhampton. Over 65 years he developed a considerable engineering empire, Firth Cleveland, including 420 shops, mainly selling radio and television sets and household appliances. His son, Jack, was a non-executive director of Firth Cleveland and had a home in the Bahamas.[39] David Owen organised a meeting between Goold-Adams and Jack Hayward. Hayward listened to the story of the ship, the plans for her rescue and restoration and at the end of the meeting, much to the relief and delight of Goold-Adams, Hayward simply said, 'That's all right. I'll see the ship home.' In the days long before such bodies as the Heritage Lottery Fund, private funders such as Jack Hayward were crucial supporters of projects. Shortly after Jack Hayward's offer to rescue the *Great Britain* the family's engineering empire was sold to GKN. The funds were now in place for the salvage.[40]

It was at this stage, with a lukewarm reception to the project in Bristol, that the ship nearly ended its days in London. Just as steam was the great revolution in shipping in the 19th century, containerisation was the revolution in the twentieth century. Container ports were having a major impact on the existing Victorian structure. The London Docks were particularly affected with large areas closed down and luxury developments in plan. St Katherine's Dock was a small dock close by the Tower of London and the Port of London Authority was looking to redevelop it. One proposal was that the *Great Britain* could be in a specialist section in the historic dock surrounded by the old warehouses. Ironically, this dock had not been a successful one, built at a time of rapid expansion and ever larger ships it never fulfilled its original purpose. Goold-Adams had several meetings with significant people in London. While Bristol was an ideal site, especially if it could be got back into its original dock, there was no disputing that a site close to the Tower of London, with its large number of visitors, might ensure the financial future of the ship.[41]

Meanwhile, fundraising continued. The organisation was still under the wing of the Society for Nautical Research with a separate account, but the *Great Britain* project now had a separate charity number from the Charity Commissioners.[42] While some of the strain was taken off the concern about fund raising with Hayward's incredibly generous offer, there were still many other things to resolve, such as who actually currently owned the ship and how the ship could be released. Ewan Corlett, meanwhile, was in negotiations with a towing company to carry out the salvage.[43]

As the profile of the project increased there was the almost inevitable misinformation and scepticism. Doubts were cast about the ship and whether it was indeed the *Great Britain*. The source of the doubt was a book written before the First World War, which stated with authority, but no evidence, that the *Great Britain* had been broken up for scrap at the turn of the century. Corlett was able to point out publicly that his measurements exactly matched the plans of the original ship and to the fact of its *Lloyd's Registry* number, which was still stamped on the side of one of the hatches and he had a photograph to prove it.[44]

From his visit, Corlett identified what he believed to a persistent and 'rooted reluctance' to destroy the ship over the course of the vessel's life in the Falklands and attributed this reluctance to his belief that the *Great Britain* had become a 'public monument' in the eyes of the Falkland Islanders. However, doubt has been cast on this by today's Falkland Islanders, who have a rather more pragmatic approach to life.[45] Lyle Cragie-Halkett, who was to become a member of the salvage team, recalled his boyhood memories of the ship as just one of the many abandoned wrecks; to him, the name Brunel meant little.

As a boy going to school I was aware of the *Great Britain* of course because we used to have annual picnics to Sparrow Cove and that's where she was but I must admit I was no more aware of the *Great Britain* than any of the other wrecks because it abounds with wrecks down there ... in the middle of the harbour there was a lovely old sailing ship called the *Fennia*. Well the *Fennia* was at anchor and you could see her every single day from everywhere you were in Stanley and of course she's swinging round at anchor and in fact we took her for granted then, but have often laughed since that the *Fennia* was used as a weather vane by all the local ladies for hanging out their washing, which line they would put it on or whatever. So you've got the *Fennia* there, a lovely big ship, not very different in size to the *Great Britain*... At the bottom of the harbour as we called it in Whalebone Cove was a ship called the *Lady Elizabeth* which is still there to this day. She's got her masts, she's got her spars and she looked absolutely beautiful in her own right. And then again you've got jetty heads also consisting of old ships, some of them much older than the SS *Great Britain* so the *Britain* was really at that time just another ship. I'd never heard of Brunel, I didn't have a clue who he was and wouldn't have been interested anyway and it really was only on going down on the job that I thought I'd better clean up my act, find out who this Brunel guy is.[46]

Some time after the salvage, Lyle Cragie-Halkett discovered that he had closer links to the ship than he thought. He was descended from one of the Scandinavian crew members from the last voyage of the *Great Britain* who had remained on the Falklands and married locally.[47]

The perception of reluctance in the Falkland Islands to part with the *Great Britain* remained an anxiety in the minds of the project team at the time. In October 1969 the project held their second press conference at the Park Lane Hotel. At this conference they announced the support of Jack Hayward, who was present, and the state of discussions with the salvage company, and finally that the *Great Britain* would 'definitely go into the Charles Hill Shipyard at Bristol'. This led to plenty of publicity, both on the television and in the newspapers and, rather aptly, a two-page article in the *Illustrated London News*, a newspaper which had carried many column inches and illustrations about the *Great Britain* since she had been built.[48]

Progress was being made on the salvage and more detailed discussions had been entered into with United Towing Company of Hull. A Dutch company had been considered but Jack Hayward was particularly concerned that the whole effort should be British. The United Towing Company wisely was not prepared to make any detailed proposals until they had seen the vessel for itself and sent a three-man team to the Falklands to look at the ship. They returned with bad news, first their view was that Falkland Island opinion was against the removal of the ship; and second, they advised against the salvage as 'the vessel did not have enough longitudinal strength left to stand up to the risk of towing the *Great Britain* in open sea.'

Corlett was staunch in his professional opinion that the salvage was possible and thoughts now turned to an alternative method so that rather than towing the ship she might be put onto another vessel. Discussions were opened with a firm in Southampton, Risdon Beazley Ltd, and they were considering the use of the technique developed by a German firm Ulrich Harms in Hamburg, which involved a submersible pontoon.[49] The steering committee of the project was now made up of Corlett, Goold-Adams, Lord Strathcona, Robert Adley MP, Custance, the treasurer, Adrian Ball, the publicist, Fitz Lombard, and Adrian Swire. Swire was an extremely well-known name in shipping circles and together with his older brother, John Swire, they brought valuable expertise.[50]

Another survey was needed and another salvage expert headed for the Falklands. With this new system of salvage, the expert pronounced the project had an 80 per cent chance of success. At last the go-ahead came from the salvage company, which was now officially called Risden Beazley Ulrich Harms Ltd. The two salvage vessels were already in West Africa and rather than returning home, they headed over to Montevideo. The tug *Varius II* carried a 21-man German team. *Mulus* was one of only three existing suitable pontoons in the world.[51]

It now required the support of the Falkland Islands Council. It was established that the ownership of the *Great Britain* had now passed from the Falkland Islands Company and, as a wreck, its disposal was now controlled by the government. This demanded a tortuous constitutional process that involved the Executive Council of the Falkland Islands, the Governor of the Falkland Islands and the Secretary of State in London, none of whom were likely to act without the support of the Falklands inhabitants.

There was still the lingering concern that the islanders had a sentimental attachment to their local landmark.[52] Part of this had come from both Corlett's visit to the Falklands and from the first salvage team's visit and it was a misunderstanding. The islanders largely viewed the ship as simply part of the landscape and one of the hundreds abandoned there. The real reluctance was on the part of the local authorities, who viewed the project with scepticism, based on previous contact, or indeed lack of contact.

At the time of the very earliest days of the project in 1967, Robert Adley, Conservative Party politician, noted railway enthusiast and then chair of the Brunel Society in Bristol, had written to Sir Cosmo Haskard, Falkland Islands Governor, in 1967.[53]

Dear Sir, As you may be aware, considerable interest has been aroused in this country recently by the plight of the Steam Ship *Great Britain* now lying in Sparrow Cove. As an admirer of Brunel, I have been discussing with friends in Bristol the possibility of investigating the condition of this famous ship. We would like to assess the likelihood of being able to bring the *Great Britain* back to her native land and restoring her. We do not underestimate the immensity of the task and in fact the first necessity is to ascertain whether or not such a journey is even feasible. This would depend quite simply on whether she could float or could be made to float without vast expenditure. We wonder whether, through your good offices, it would be possible to find someone locally to give us a reasonably accurate report as to the actual condition of the *Great Britain*, and we hope we might arouse in the Islands some enthusiasm for the task.[54]

At the time, the Islands administration knew little of the plans being formulated in Britain. They did, however, know of the San Franscico project and its plans. Karl Kortum from San Francisco had at least visited the ship in 1967 and his project was seen as more serious.[55]

Responding to Adley's letter, W. H. Thompson, the Colonial Secretary of the Falkland Islands in 1967, was less than positive about the *Great Britain* returning home.

I would be doing you a wrong if I encouraged you to believe that the hulk is something which a few enthusiasts might float away and restore. In my last letter I gave you a figure of a million pounds as a possible figure. Since writing my thoughts have tended towards something much larger.[56]

In 1968 Thompson wrote to the Commonwealth Office in London and made a distinction between the two bids to salvage the ship: 'A representative from the San Francisco Maritime museum has visited the Falkland Islands and has seen the remains. No representative of any United Kingdom organisation has seen the ship.'[57] Corlett did not go out until 1968 and it is this scepticism he translated into the reluctance of the islanders to lose the ship. In the event, the San Francisco Museum turned their thoughts to another ship in the Falkland Islands, the *Fennia*.[58] This was a Finnish-owned vessel, a four-masted steel barque built in 1902, that had been dismasted and left in the Falkland Islands. Bought by the San Francisco Museum, it was towed to Montevideo, but the project ran out of funds and the ship was eventually taken to Paysandu, Uruguay, for scrap.[59] This failed salvage did not help views of the likelihood of success for the British project.

The Commonwealth Office wrote to the governor with a strong hint that it would look forward to receiving his recommendation for the release of the *Great Britain*, and the Executive Council of the Falkland Islands finally agreed that the ship could be released. But there were conditions, influenced by concern for shipping.

A. Release should be only for pontoon method and not some other method
B. Any interference with shipping channel leading to Stanley harbour which might result from salvage or towing would be remedied at the expense of the committee and if any point to the pontoon method seeming unlikely to succeed operation should be suspended before too much damage had been done to the old ship.[60]

Before the vessel could be officially handed over there was the slight anomaly to be resolved in the status of the receiving organisation. It was still effectively just a committee, albeit recognised as an affiliated body, a subsidiary of the Society for Nautical Research.[61]

This point was eventually overcome and the necessary documents were secured. The rescue of the ship could go ahead.

The success of the project and its continued determination to remove all obstacles was remarkable. It was an unusually difficult matter when considering the 7,000 miles between the *Great Britain* and the project team ship, who were attempting a complex industrial archaeological project at a distance. Before internet and easy telephone connections, much of the correspondence, principally with the Falkland Islands administration, was by letter.[62] Corlett, Strathcona, Goold-Adams and Hayward saw the *Great Britain* as a symbol of Great Britain at its industrial height.

Corlett remained deeply inspired by the life and achievements of Brunel.[63] In 1976 he wrote:

Everything I have done in connection of the *Great Britain* has increased my respect for Brunel's foresight, engineering ability and imagination. ... with this time [when] national self-confidence is at a low ebb it is well worth reading what can be done with imagination.[64]

By the autumn of 1969 the scheme was coming together and the newspapers were publicising their efforts. *The Times* explained the planned rescue describing the ship as a 'unique nautical antique and a memorial of the great days of Victorian shipbuilding', a symbol of the transition from sail to steam and pictured it 'beached at Sparrow Cove, in the Falklands, scoured by tides that threaten to break her back within five years'. Jack Hayward, 'that enthusiastic Patriot', was asked why he had volunteered the money. His reply was 'because she belongs to us. She bears our name- the *Great Britain*. She has no right to go anywhere except her homeland. And if we let her go anywhere else, we deserved to be a second-class power.' The ship would be brought back to complete her last voyage, voyage number 47, so 'rudely interrupted by storm off Cape Horn 84 years ago'. And when she was safely back she would be restored 'down to the last loving detail of Brunel's revolutionary blueprints. This is like to take two years and cost up to £250,000.'[65]

Euan Strathcona would be the only 'project representative' present at the actual salvage and would be able to take formal release of the ship from the governor of Falkland Islands. He took with him his brother-in-law, Viscount Chewton. They had secured press accreditation respectively for the *Daily Telegraph* and from *The Times*. Ray Sutcliffe, the producer of BBC 2's programme, *Chronicle*, was accompanied by Tony and Marion Morrison, an experienced camera team.[66]

Strathcona was shocked when he first climbed aboard the vessel.

Well I was pretty amazed and when we went on board I was horrified to find that where the crack was you could stand on it and you could feel the ship moving on either side of the crack. The importance of that is that, as Ewan Corlett pointed out, probably in a matter of months, certainly not very many years, the ship would have broken in half and that would have made the salvage job a lot more difficult if not actually impossible.[67]

British Government support was crucial to the eventual success of the salvage. Work began on 26 March and yet again there was extensive help from the Navy. Royal Marines transported the salvage team between Stanley and Sparrow Cove by hovercraft. They assisted in removing the masts and fitting planks. The salvage team bolted steel straps across the large crack and the drilling to secure these was anything but easy, proving the strength remaining in the ship. Patches covered the holes that had been made when the ship was scuttled. An old method was now used to fill the crack and requests went out to anyone who might supply redundant mattresses. The navy yet again helped with the task of sourcing and transporting them to fill the hole in the *Great Britain's* hull before the ship was refloated. Hans Herzog, Captain of the *Varius* (the tug that would eventually tow the *Great Britain* back to Bristol on top of the floating pontoon, the *Mulus*) saw this support as vital to the success.[68]

There was also local help from Lyle Cragie-Halkett, who was both a diver with the salvage team and a Falkland Islander. Looking back he was very proud of his professional part in the salvage.[69] Interviewed in 2016, the pride in which Falkland Islanders spoke of their own involvement with the salvage of the *Great Britain* suggests the controversy relating to the ship's removal from the Falkland Islands was 'potentially limited in its extent and certainly not a source of continued debate'.[70]

Now came the tricky task of getting the ship onto the pontoon and the March weather in the Falklands was not helping, it was blowing a gale. On 5 April the job of pumping out the ship began. The pontoon was submerged and moved underneath the ship. At last on 14 April the ship was wedged firmly on the pontoon and towed by tug from Sparrow Cove to Stanley. Here with due ceremony the Falkland Islands government handed over formal ownership of the *Great Britain* to Lord Strathcona. The ship was 127 years old and had been beached as a wreck for 33 years. She now began the final part of the very last voyage. She went into Montevideo to a great welcome and sailed from there on 6 May. She then headed out to sea again for her long, risky transatlantic passage. Sudden severe weather could have endangered the tow, but it went well and the weather was kind until four days of poor weather off the Canary Islands, when the tug could make little headway. As the convoy entered the Bay of Biscay, an RAF Nimrod circled the ship. On board was a press photographer and the resulting picture was published in the *Daily Express* with the simple caption 'Grandmother of them all'.[71]

On 24 June the pontoon with its precious load entered Avonmouth docks in Bristol to be met by the project team, who had followed her progress with a mix of pride and trepidation.[72] Crowds and noise greeted her as tugs eased her through the lock and up to the dockside and the reporter from *The Times* captured the excitement of the committee, most of whom were seeing the ship for the very first time.

> Sirens Greet Return of Rusty *Great Britain*: One of the great maritime relics of all time, the steamship *Great Britain*, all rust and barnacles, was brought back to England today from her misty Falklands grave. The men who invested the money, the faith and the time to achieve the unlikely and epic feat of salvaging the ship and bring her home, clambered round their prize like boys around a steam engine.[73]

Jack Hayward had flown from the Bahamas to see the ship come home. 'With a broad smile he toured the decks with Dr Ewan Corlett, the naval architect whose letter to *The Times* three years ago started the salvage effort.'[74] Not everyone had the same romantic view of the ship and many wondered what could be made of the wrecked mass of iron rust. 'The lower part of the hull is covered with marine growth; inside she is a mass of rusty debris.'[75] And yet, as many letters to the newspapers attested, Bristolians now saw the ship as part of the city's heritage and history.[76]

On 5 July 1970 the ship freed from its pontoon was towed up the river Avon to Bristol. The committee had taken out a £2 million insurance policy to cover the risks of the delicate passage nine miles upriver.[77] This was a massive event, heavily televised across the nation. Tens of thousands of people lined the banks and it was a particularly historic moment when she was towed under Brunel's completed Clifton Suspension Bridge, a bridge that had existed only in Brunel's plans at the time of the first passage of his ship down the Avon. That morning the *Great Britain* floated safely through the gates back into the floating harbour to await her return to the Great Western Dock.[78]

On 19 July 1970, on the anniversary of her original floating out, the *Great Britain* returned to her original dock that she had left exactly 127 years before. There she was greeted by great crowds including His Royal Highness Prince Philip, Duke of Edinburgh, in the same way that his predecessor, Prince Albert, had waved out the ship in the last century. It was an incredible feat of determination by a small group of enthusiasts to bring this amazing ship back to its home.

The work had only just begun. The next stage was to restore her to a ship that Brunel, Guppy, Claxton and Patterson would recognise; but not before many toasts had been drunk in celebration of her return home. In 1970 in just seven weeks the rusting hulk attracted 100,000 visitors, proving its incredible popularity at such an early stage.[79]

The fact that we are still able to visit the ship and admire her is a fitting tribute to her workmanship and design overseen by Isambard Kingdom Brunel. The *Great Britain* led the way in propeller technology and the use of iron, promoting the use of the latter even when she was stranded. That the *Great Britain* has survived through lengthy service across the Atlantic, the Indian and Pacific Oceans, around Cape Horn and abandonment for so many years in the Falklands is a testament to her building committee. Brunel's work for the navy as a consultant on trials of HMS *Polphemus* and the design of HMS *Rattler* ensured the commissioning of propeller-driven warships. The *Great Britain* influenced the first iron warship, *Warrior*.

The *Great Britain* was, and is, unique. In 1863, the noted Victorian shipbuilder, Scott Russell, looked back at the *Great Britain* and pointed out that he had been consulted by the Great Western Steam Ship Company at an early stage and he had advised that the ship should be somewhat larger than the *Great Western* but be very similar in design. The company decided not to do that but, as Scott Russell described it, they opted for a whole design 'entirely revolutionised, and turned into an imitation of Sir W. Symond's new and empirical form of ship'. In Scott Russell's opinion every feature of the vessel's design was experimental and the ship itself was unique.[80] The reference to Symonds is not wholly flattering. Sir William Symonds was the surveyor to the navy from 1832 to 1847 at the time of the screw propeller trials. His ship designs were ill suited to the installation of machinery.[81]

This uniqueness was due to Brunel and his colleagues in the Great Western Steam Ship Company, who had an engineering ambition far beyond day-to-day commercial concerns. The *Great Britain* was conceived and built in Bristol, with shareholders whose dream was to make Bristol a rival port to Liverpool. The company had visions of a fleet of innovative ships and planned to build ships for other companies. These ambitions were unfulfilled and the lack of wider funding for the company led to its failure as a business, together with the realisation of the limitations of Bristol as a port suitable for large ships.

The *Great Britain* has also been a very lucky ship, despite the Dundrum Bay incident. The Australia opportunity came at the right time, just as the Great Western Steam Ship Company was desperate to sell her. Gibbs, Bright proved to be shrewd and effective owners. Her abandonment in the Falkland Islands saved her for posterity, had she been left in Britain she would have been broken up, as was the fate of the *Great Eastern* and multitudes of other iron ships.

She is a ship that engenders strong feelings; witness Brunel's passionate outburst at Dundrum Bay, Captain Gray's love of the ship, Henniker Heaton's efforts that saved the ship from destruction and the unremitting work of Corlett, Goold-Adams,

Hayward and the rest of the committee of the rescue project. Even the Falkland Islanders now view her with pride.[82] Today she continues to attract praise. Restored and cared for by the staff and volunteers over the years, her attractive looks, her elegant design as she floats above her glass sea in her dock, mean she fully deserves the many awards she and the SS Great Britain Trust have collected. We are deeply indebted to those first volunteers whose passion ensured she returned to Britain.

Appendix 1

TIMELINE

1837	Launch of *Great Western*
1838	Purchase of timber for ships, Purchase of new dock
	Arrival of *Rainbow*, iron ship
1839	Directors announce new ship will be iron
	Quotes for engines
(July)	Fabrication begins
(Dec.)	Keel laid
1840	Arrival of *Archimedes* propeller-driven ship
	Brunel sketches propeller designs
(Dec.)	December decision to go with propeller
	Guppy appointed as Directing Engineer, Harman as engineer in chief.
1841	AGM ship now due March 1842
	Colonial Land and Emigration Commission established
1842	AGM Ship now named *Great Britain* and delayed to spring 1843
(July)	Dockyard lease for sale
(Nov.)	Extraordinary general meeting
	Letter from shareholders
1843	*Great Western* sailing permanently from Liverpool
	HMS *Rattler* Royal Navy
(July)	Floating out/Launch of *Great Britain*
	Guppy registers patent for iron lifeboats
1845	Brunel drawing of ship with two paddlewheels

1845	First paying voyage Liverpool to New York
1846	*Great Britain* stranded in Dundrum Bay
1847	Sale of *Great Western* to Royal Mail Steam Packet Company
1847	*Great Britain* salvaged
1850	Sale of *Great Britain* to Gibbs Bright
1851–53	Brunel, consulting engineer for Australian Royal Mail Steam Ship Company
1852	*Adelaide* and *Victoria* for Australian Royal Mail Steam Ship Company
1852	Drawings of *Wave Queen*, *Victoria*, *Great Britain* and unnamed vessel
	Gold rush in Australia
(May)	*Great Britain* Liverpool to New York after refit
(Aug.)	*Great Britain* to Australia
1853–58	The Great Ship project commences for Eastern Steam Navigation Company
1854	Letter to Brunel from Claxton about *Great Britain*
(Sep.)	SAllied invasion of Crimean peninsula
1855	March *Great Britain* Troopship in Crimea
	Launch of *Royal Charter*
1857	post-Crimea refit
(Feb.)	to Australia
(Sep.)	Trooping run to Bombay
1858	Launch of *Great Eastern*
(July)	*Great Britain* from Liverpool and New York
(Nov.)	To Melbourne
1859	Last visit of *Great Britain* to New York
(Sep.)	Death of Brunel
1859	Loss of the *Royal Charter*
1868	Claxton dies
1869	Patterson dies
1869	Opening of Suez Canal
1872	Disappearance of Captain Gray
1873	The Anglo-Australian Steam Navigation Company Limited announced
(Oct.)	*Great Britain* goes to Melbourne and Brisbane
1876	February final Australia voyage

1881	Gibbs Bright now Anthony Gibbs
1882	Sale of contents of *Great Britain*
	Refit by Graysons and converted to sailing cargo ship
	Death of Guppy in Naples
(Nov.)	*Great Britain* sails from Liverpool to San Francisco
1884	May sails second time to San Francisco
1886	Sails from Penarth to San Francisco with coal
(May)	*Great Britain* seeks shelter in Port Stanley
1887	Sold to Falkland Islands Company
1936	Henniker Heaton tries to save *Great Britain*
1967	Corlett writes to *The Times*
1968	Meeting of Corlett, Greenhill and Goold-Adams with Falkland Islands Company
(Sep.)	Corlett surveys the *Great Britain* in Falkland Islands
1969	Jack Hayward agrees to fund return
1970	Start of savage operation 24 June
(Apr.)	*Great Britain* leaves from Sparrow Cove
(June)	Triumphant return to Bristol
(July)	*Great Britain* back in the Great Western Dock

APPENDIX 2

VOYAGES OF THE
GREAT BRITAIN

No	Date		Arr	Master
1	Jan 1845	Trials and voyage to London		Hosken
2	May 1845	London to Liverpool		Hosken
3	26.07.1845	Liverpool–New York–Liverpool	15.09.1845	Hosken
4	27.09.1845	Liverpool–New York–Liverpool	17.11.1845	Hosken
5	09.05.1846	Liverpool–New York–Liverpool	21.06.1846	Hosken
6	07.07.1846	Liverpool–New York–Liverpool	15.08.1846	Hosken
7	22.09.1846	Liverpool–stranded at Dundrum Bay	22.09.1846	Hosken
8	01.05.1852	Liverpool–New York–Liverpool	16.06.1852	Mathews
9	21.08.1852	Liverp–Melbourne–Sydney–Melbourne–Liverp	02.04.1853	Mathews
10	11.08.1853	Liverp–Melbourne–Sydney–Melbourne–Liverp	14.02.1854	Mathews
11	12.06.1854	Liverpool-Melbourne-Liverpool	24.01.1855	Gray
12a	07.03.1855	Crimean War troopship	14.07.1856	Gray
12b	14. 04.1855	Crimean War troopship	16.8.1855	Gray
12c	04. 09.1855	Crimean War troopship	25 10 1855	Gray

No	Date		Arr	Master
12d	17. 11.1855	Crimean War troopship	18.12.1855	Gray
12e	02. 01.1856	Crimean War troopship	29.05 1856	Gray
13	16.02.1857	Liverpool–Melbourne–Liverpool	22.08.1857	Gray
14	24.09.1857	Cork–Bombay–Liverpool	10.04.1858	Gray
15	28.07.1858	Liverpool–New York–Liverpool	07.09.1858	Gray
16	21.11.1858	Liverpool–Melbourne–Liverpool	08.05.1859	Gray
17	01.07.1859	Liverpool–New York–Liverpool	10.08.1859	Gray
18	11.12.1859	Liverpool–Melbourne–Liverpool	07.05.1860	Gray
19	20.07.1860	Liverpool–Melbourne–Liverpool	01.01.1861	Gray
19	20.07.1860	Liverpool–Melbourne–Liverpool	01.01.1861	Gray
20	17.02.1861	Liverpool–Melbourne–Liverpool	04.08.1861	Gray
21	20.10.1861	Liverpool–Melbourne–Liverpool	08.04.1862	Gray
22	15.06.1862	Liverpool–Melbourne–Liverpool	20.11.1862	Gray
23	25.01.1863	Liverpool–Melbourne–Liverpool	03.07.1863	Gray
24	15.10.1863	Liverpool–Melbourne–Liverpool	06.04.1864	Gray
25	26.05.1864	Liverpool–Melbourne–Liverpool	28.10.1864	Gray
26	17.12.1864	Liverpool–Melbourne–Liverpool	20.05.1865	Gray
27	25.07.1865	Liverpool–Melbourne–Liverpool	22.12.1865	Gray
28	18.02.1866	Liverpool–Melbourne–Liverpool	20.07.1866	Gray
29	27.10.1866	Liverpool–Melbourne–Liverpool	29.03.1867	Gray
30	19.05.1867	Liverpool–Melbourne–Liverpool	19.10.1867	Gray
31	15.12.1867	Liverpool–Melbourne–Liverpool	23.05.1868	Gray
32	09.07.1868	Liverpool-Melbourne-Liverpool	07.12.1868	Gray
33	03.02.1869	Liverpool–Melbourne–Liverpool	09.07.1869	Gray
34	12.08.1869	Liverpool–Melbourne–Liverpool	04.02.1870	Gray
35	19.03.1870	Liverpool–Melbourne–Liverpool	23.08.1870	Gray
36	06.10.1870	Liverpool–Melbourne–Liverpool	22.03.1871	Gray

No	Date		Arr	Master
37	24.05.1871	Liverpool–Melbourne–Liverpool	08.11.1871	Gray
38	17.12.1871	Liverpool–Melbourne–Liverpool	28.05.1872	Gray
39	27.07.1872	Liverpool–Melbourne–Liverpool	25.12.1872	Gray
40	30.03.1873	Liverpool–Melbourne–Liverpool	03.09.1873	Chapman
41	27.10.1873	Liverp–Melboune–Brisbane–Melbourne–Liverp	17.04.1874	Chapman
42	04.06.1874	Liverpool–Melbourne–Liverpool	18.11.1874	Chapman
43	11.01.1875	Liverpool–Melbourne–Liverpool	10.07.1875	Chapman
43	11.01.1875	Liverpool–Melbourne–Liverpool	10.07.1875	Chapman
44	26.08.1875	London–Melbourne–Liverpool	30.01.1876	Robertson
45	02.11.1882	Liverpool–San Francisco–Liverpool	01.02.1884	Stap
46	11.05.1884	Liverpool–San Francisco–Liverpool	08.07.1885	Stap
47	06.02.1886	Penarth (Cardiff) – stranded in Falklands	24.05.1886	Stap

Ewan Corlett, *The Iron Ship: The Story of Brunel's SS Great Britain*. 2012 edn (Bristol: ss Great Britain, 1975), pp. 328-34

APPENDIX 3

THE GREAT WESTERN STEAM SHIP COMPANY AND LIMITED LIABILITY

To understand the Great Western Steam Ship Company it is necessary to understand the relationship between the directors and the proprietors. The directors always referred to themselves as proprietors, indicating a close ownership, while the investors referred to themselves as shareholders.

A business partnership was unlimited as each partner was wholly liable for the actions of the other so all their property was at risk, regardless of any investment. In shipping there was a form of limited liability in the 64th system of ship ownership, whereby each shipowner was only liable for their shares in the ship and it came under Admiralty law jurisdiction. The Great Western Steam Ship Company with its ambitious plans avoided both the partnership model and the 64th system and chose a more complex form.

The choice for entrepreneurs requiring large amounts of capital was either to become an incorporated joint stock company via an Act of Parliament, a lengthy and costly process, or form a 'voluntary unchartered shareholding partnership', which combined trust, partnership, equity, contract and agency law. The company chose this route, an unincorporated joint-stock company set up by deed of settlement (AGM 1837). A wide range of companies from canals, docks, fire and life insurance, bridges, fisheries, brewing, timber trade and shipping created such companies and from 1810 to 1840 the growth was considerable. This form of company dominated business organisations up to the 1840s.

An unincorporated group of individuals could not hold or own property as a group, the property could be held in trust for it. This was achieved by means of mutual covenants between the shareholders of the company and the trustees selected by them. These were set out in a company's deed of settlement, and the trustees undertook to observe the terms of the deed and to use the company's funds only for the purposes specified. The company was therefore unable to act through its trustees rather than as a mass of individuals, approximating the corporation's ability to sue and be sued in its own name. Lawyers who drew up the deeds also attempted to make the company's shares freely transferable and in some cases attempted to limit the liability of shareholders. [Taylor, *Creating Capitalism*, p 4]

The challenge in any unincorporated stock company was that they lacked a legal entity and ownership and trading of company shares were not recognised at law. To enhance their legitimacy unincorporated stock companies often identified the company with a particular locality and region and canvassed support of leading patrons with overlapping business partnerships and family connections, and involvement in local political and economic institutions. This was particularly the case in insurance companies. It was not until 1844 that such companies could be registered and by 1856 there was general limited liability.

Liabilities of shareholders remained a problem, so they appointed trustees to hold the company assets and in the case of the Great Western Steam Ship Company there were four trustees. It is in their names that the ship was registered. In the deed of settlement it was common to limit the liability of individual partners for the debts of the firm to the value of each partner's investment. Many shares in unincorporated companies were circulated, even though this was forbidden.

In the dispute with the Bristol Dock Company they had legal advice from T. Tomlinson at Temple. On 14 February 1844 regarding the contract and the Bristol Dock Company, he refers to the directors of the Great Western Steam Ship Company and their deed of settlement. 'But as the company is not protected by any charter or Act of Parliament, the directors can only bind its funds and capital stock through the medium of the personal liability of the partners, and must, therefore, at least, subject themselves to personal responsibility under any contract which they may enter into.' (Great Western Steamship Company AGM March 1844)

One of the resolutions in 1844 was that the directors be requested to 'consider and propose a plan for the division of the present shares of the company, in such a manner as to assign to each proprietor separate and distinct shares in the ships and the buildings of the company, upon an equitable valuation of the whole property.' (Great Western Steamship Company AGM March 1844) This would be an attempt to limit each shareholder's liability. Until the Limited Liability Act of 1855 this was a very grey area and highlights the risk for the company's shareholders.

Sources

Robin Pearson, 'Shareholder Democracies? English Stock Companies and the Politics of Corporate Governance During the Industrial Revolution', *The English Historical Review*, 117 (2002) p. 845-50

James Taylor, *Creating Capitalism: Joint Stock Enterprise in British Politics and Culture, 1800-1870* (London: Royal Historical Society, 2006) p. 4

BI: Great Western Steamship Company AGM papers

APPENDIX 4

THE GREAT WESTERN STEAM SHIP COMPANY SHAREHOLDERS/PROPRIETORS

No complete list of the shareholders has emerged. What evidence there is comes from AGM reports and newspaper reports. The following is a list of names and roles as shown in these. Only the names mentioned are listed, so, for example, Claxton and Guppy certainly were directors for almost the whole time. Most of the shareholders were Bristol or Bath residents, with the former being the larger group. There was one lady shareholder mentioned.

1839	Peter	Maze	Chairman
1839	Thomas	Kington	deputy chairman
1839	Robert	Bright	Director
1839	Henry	Bush	Director
1839	Henry	Godwin	Director
1839	Thomas R	Guppy	Director
1839	Robert	Scott	Director
1839	Thomas Boneville	Were	Director
1839	Christopher	Claxton	managing director
1839	Joseph	Cookson	Trustee
1839	John	Harford	Trustee

1839	Thomas	Kingsbury	Trustee
1839	John	Vining	Trustee
1839	Charles Bowles	Fripp	auditor
1839	John	Moxham	auditor
1839	John	Winwood	auditor
1839	W H	Castle	Shareholder/proprietor
1839	Thomas	Cruttwell	Shareholder/proprietor
1839	T H	England	Shareholder/proprietor
1839	F H	Falkner	Shareholder/proprietor
1839	Dr Charles	Fox	Shareholder/proprietor
1839	Thomas	Kingsbury	Shareholder/proprietor
1839	Samuel	Lucas	Shareholder/proprietor
1839	Miss	Lucas	Shareholder/proprietor
1839	William	Morgan	Shareholder/proprietor
1839	Richard	Robinson	Shareholder/proprietor
1839	John	Stafford	Shareholder/proprietor
1839	Captain Edward	Walcott RN	Shareholder/proprietor
1839		Woodman	Shareholder/proprietor
1839	Miles, Harford & Co, Bristol		Bankers
1839	Barnetts, Hoare & Co, London		Bankers
1839	Osborne, Ward & Sons, Bristol		Solicitors
1839	Swain, Stevens & Co, London		Solicitors
1840	Peter	Maze	Chairman
1840	Thomas	Kington	Director
1840	Robert	Bright	Director
1840	Henry	Bush	Director
1840	Thomas Boneville	Were	Director
1840	Robert	Scott	Director
1840	Thomas R	Guppy	Director

1840	Henry	Godwin	Director
1840	Christopher	Claxton	Director
1840	J B	Clarke	Shareholder/proprietor
1840	Capt	Crozier RN	Shareholder/proprietor
1840	Charles Bowles	Fripp	Shareholder/proprietor
1840	George	Gibbs	Shareholder/proprietor
1840	John	Hare	Shareholder/proprietor
1840	John	Harford	Shareholder/proprietor
1840		Levy	Shareholder/proprietor
1840	John	Moxham	Shareholder/proprietor
1840	Jere	Osborne	Shareholder/proprietor
1840	Robert	Osborne	Shareholder/proprietor
1840	Thomas	Pycroft	Shareholder/proprietor
1840		Smith	Shareholder/proprietor
1840		Stedder	Shareholder/proprietor
1840	R B	Ward	Shareholder/proprietor
1841	Robert	Bright	Chairman
1841	Peter	Maze	stepped down
1841	Henry	Bush	Director
1841	John William	Miles	Director
1841	Thomas Boneville	Were	Director
1841	Christopher	Claxton	Director
1841	J P	Bartrum	Shareholder/proprietor
1841	J	Bates	Shareholder/proprietor
1841	M H	Castle	Shareholder/proprietor
1841	Joseph	Cookson	Shareholder/proprietor
1841		Cunningham	Shareholder/proprietor
1841	A P	Johnson	Shareholder/proprietor
1841	George	Jones	Shareholder/proprietor
1841	William	Morgan	Shareholder/proprietor
1841	George	Norman	Shareholder/proprietor
1841	Jere	Osborne	Shareholder/proprietor
1841	Thomas	Pycroft	Shareholder/proprietor

1841	Charles Bowles	Fripp	auditor
1842	Thomas	Kington	Chairman
1842	Robert	Bright	Director
1842	Henry	Bush	Director
1842	Thomas Boneville	Were	Director
1842	John William	Miles	Director
1842	Thomas R	Guppy	Director
1842	Henry	Godwin	Director
1842	Christopher	Claxton	Director
1842	Robert	Scott	
1842	Thomas	Pycroft	
1842	George	Gibbs	auditor
1843	Thomas	Kington	Chairman
1843	Robert	Bright	deputy chairman
1843	Henry	Bush	Director
1843	Thomas Boneville	Were	Director
1843	John William	Miles	Director
1843	Thomas R	Guppy	Director
1843	Henry	Godwin	Director
1843	Christopher	Claxton	Director
1843	Adam	Holden	auditor
1843	Robert	Scott	steps down as Director
1843	Thomas	Pycroft	Director according to news report
1843	J	England	Director according to news report
1843	R B	Ward	Director according to news report
1844	Henry	Bush	Chairman
1844	Thomas	Kington	chairman
1844	Peter	Maze	chairman
1844	Henry	Godwin	Director
1844	Charles Bowles	Fripp	auditor
1848	Henry	Bush	Chairman

1848	John William	Miles	Director
1848	Thomas	Kington	Director
1848	Thomas Boneville	Were	deceased
1848	George	Gibbs	auditor
1848	Charles Bowles	Fripp	auditor
1848	George	Norman	Shareholder/proprietor
1848		Abbot	
1848	Henry	Godwin	
1848	Joseph	Cookson	
1848	John	Taylor	
1848		Walton	
1848		Hollister	

Sources: BI: Great Western Steam Ship Company AGM papers; *Bristol Times & Mirror* 22 July 1843 & 4 March 1848

APPENDIX 5

IN DEFENCE OF CAPTAIN HOSKEN

Hampshire Advertiser, Saturday 24 October 1846

PORTSMOUTH, OCT. 24, 1846. The character of Lieutenant Hosken has long stood too high, both in the Navy and in the Merchant Steam-service, for any one conversant with either to suspect that the unfortunate loss of the *Great Britain* was in any way caused by want of skill or care on the part of her commander; but as some other persons speak of it in a way to be excused only by their ignorance of the circumstances of the case, it may not be too late to call their attention to the main fact at least. The *Great Britain*, then, ran ashore simply because the St. John's Light, on the coast of Ireland, not being inserted in the charts belonging to the ill-fated steamer, was mistaken for that on the Calf of Man.

As early as the beginning of this month, a naval friend wrote to us – "The fault seems to have been the omission of a light-house in a brand-new chart. The man ought to be stuck on the top of the light-house and made to burn a blue light, every fine minutes of the night, for the remainder of the winter, as a punishment. I know Hosken well, and am persuaded he is not the man to do a careless or a lubberly thing." The character of our informant led us to rely at once on both his assertions; the former has since been confirmed by statements in various papers, and the latter can well take care of itself. A writer in the Standard having laid the blame both of the grievous error in question, and of another, on the Admiralty charts, Mr Bate, the agent for their sale, states that the chart used by Captain Hosken is not an Admiralty, but a Liverpool one, "which though sold by his agent, at Liverpool, has no authority on that account." The other charge, which he also rebuts, concerns an error which might prove equally fatal – namely, the placing of a light-house in the Murray Firth, a part of the Scotch coast of which, it seems, there is no Admiralty chart (why is there not ?); whether the incorrect one complained of by the correspondent of the Standard is also a Liverpool production, and sold there, or elsewhere, by the agent for the Admiralty charts, does not appear. We rejoice that our official hydrographers are thus cleared from two such serious accusations; while, however, the Admiralty cannot of course, be held responsible for any but their own charts their agents might, we think, be restrained from selling any others not inspected and approved by the hydrographic department.

Any one buying charts of such a person without actually supposing them to "have authority on that account," might be excused for thinking them more likely to be tolerably correct, than if procured elsewhere. Anyhow, the imputation of carelessness or ignorance being thus removed from an excellent officer, should be fixed on a bad chart-maker. The fate of the *Great Britain* suggests more than one train of thought, which cannot here be followed out.

One thing is certain, that far worse might have happened; as many, and those not inexperienced persons, felt serious uneasiness as to how a vessel of such enormous length, however well built, might meet an Atlantic gale – like, for instance, that which the *Great Western*, a ship of most approved sea-going qualities and power, has hardly weathered. That the catastrophe, if come it must, took place where and how it did, should give cause for special thankfulness.

Without any wish to interpret judgments, one cannot help recalling with a feeling of pain the boastful tone in which the unparalleled size, and other remarkable characteristics of this iron leviathan, were spoken of both by those interested in her and by others who seem to have allowed their calmer judgment to be overwhelmed and swept away by the tide of vulgar admiration for vast bulk, and portentous novelty. To take the lowest ground, the prudent always condemned the speculation as a most perilous one; had the Company been content with another *Great Western*, their loss, anyhow, would have been far smaller, and their fall far less humiliating. An iron-screw steamer of 1,500, or at most 2,000 tons, might, one would think, have afforded sufficient novelty to satisfy that spirit of enterprise and experiment which has been pleaded in excuse both for the enormous dimensions and enormous risk of such a vessel, and for the unseemly self-congratulations and glorifications with which she has been celebrated.

Both the early success, and the final fate of the *Great Britain*, will confirm the daily-growing opinion in favour of iron as a material, and of the screw as a mean of propulsion. With regard to the former, hardly any wooden vessel – none certainly of equal length (unless perhaps built on that plank-layer system, which after successful trials in the merchant service is now most unaccountably confined to the boats of men-of-war) – would have held together as she has done; this point, however, had been already pretty well proved by the example of the more fortunate *Nemesis*, which was bumped about on the rocks of China till her bottom was as full of dents and scratches as that of an old tea-kettle. The triumphant success of Mr Smith's screw-propeller has never been doubtful since the first trial of the beautiful *Archimedes*, whose appearance in these waters was to us the presage of a new generation of ships, which, combining all the elegance and ship-shape of the old sailing-vessel, with far more than all the advantages of the newer paddle-wheel steamer, must, ere long, take the place of both, as every way fitter for carrying to the ends of the earth the many who shall pass to and fro, and the knowledge which shall be increased.

APPENDIX 6

CAPTAIN CLAXTON'S HOUSEHOLD SALE, 1849

Bristol Times and Mirror, Saturday 10 March 1849

Valuable effects St Vincents Parade
Messrs Fargus & Son
Are instructed to submit to Public competition, on the premises, on
Thursday and Friday, the 22nd and 23rd of March

All the handsome household furniture China glass
Large brilliant cut glass chandelier with 12 lights
A fine toned grand six ½ octaves pianoforte by Broadwood etc
Circular library table
Two excellent double barrel guns and cases one by Mortimer
and other valuable effects of Captain Claxton RN leaving the
neighbourhood

Comprising a suit of fashionable figured drab damask drawing-room
curtains, with gilt cornices: sofas, couches, and chairs, in needle
work cases with figured satin covers over loo, card, and occasional
tables; a pair of rosewood tables, on gilt tripods and turtle stone
tops; excellent Brussels and other carpets; mahogany four post,
tent, and other bedsteads with chintz furniture; prime feather and
mill puff beds; mattresses and bedding; neat mahogany wardrobes;
chests of drawers; bookcases; dressing tables and glasses; wash

stands, with marble tops; set of mahogany dining tables, on claws; sideboard, with cellarette; mahogany dining and lounging chairs; eight day and other clocks; Ruby moreen dining parlour curtains; whatnots; hat and umbrella stands; several large and small cupboards and linen presses; a great variety of kitchen requisites, and other useful articles.

The Sale will commence each Morning precisely at Eleven O'Clock, and the whole may viewed the day preceding the Sale.

APPENDIX 7

EMIGRANT REGULATIONS (ORDERS IN COUNCIL) 1852

Morning Post, Wednesday 27 October 1852

EMIGRANT SHIPS. The Gazette of last night contains the following Order in Council, for the preservation of the health of passengers proceeding from the United Kingdom to any port or place in her Majesty's possessions abroad: – At the Court at Windsor, the 16th day of October, 1852, present, the Queen's Most Excellent Majesty in Council. Whereas by an act, passed in the session of Parliament held in the 15th and 16th years of the reign of her Majesty: intituled the "Passengers Act, 1852," it is enacted, that it shall be lawful for her Majesty, by any order in Council to be by her made, with the advice of the Privy Council, to prescribe such rules and regulations as to her Majesty may seem fit, for preserving order, for promoting health, and for securing cleanliness and ventilation, on board of passenger ships proceeding from the United Kingdom to any port or place in her Majesty's possessions abroad, and the said rules and regulations from time to time in like manner to alter, amend, and revoke, as occasion may require: And whereas it is expedient to revoke an order in Council, made at a Court, held at Osborne House, Isle of Wight, on the 6th day of October, 1849, in virtue of the provisions of the Passengers Act, 1849 (now repealed), and to make a new order in Council: Now, therefore, her Majesty doth, by and with the advice of her Privy Council, and in pursuance and exercise of the authority vested in her by the said Passengers Act, 1852, order, and it is hereby ordered, that the said order in Council of the 6th day of October, 1849, be, and the same is hereby revoked; and that the following shall henceforth be the rules for preserving order, for promoting health, and for securing cleanliness and ventilation, to be observed on board of every passenger ship proceeding from the United Kingdom to any port or place in her Majesty's possessions abroad out of Europe, and not being within the Mediterranean Sea.

1. All passengers who shall not be prevented by sickness or other sufficient cause, to be determined by the surgeon, or in ships carrying no surgeon, by the master, shall rise not later than seven o'clock a.m., at which hour the fires shall be lighted.
2. It shall be the duty of the cook or cooks, appointed under the 36th section of the said Passengers Act, 1852, to light the fires, and to take care that they be kept alight during the day; and also to take care that each passenger or family of passengers shall have the use of the fireplace at the proper hours, in an order to be fixed by the master.
3. When the passengers are dressed, their beds shall be rolled up.
4. The decks, including the space under the bottom of the berths, shall be swept before breakfast, and all dirt thrown overboard.
5. The breakfast hour shall be from eight to nine o'clock, a.m. Before the commencement of breakfast, all the emigrants, except as hereinbefore excepted, are to be out of bed and dressed, and the beds rolled up, and the deck on which the emigrants live properly swept.
6. The deck shall further be swept after every meal, and after breakfast is concluded, shall be also dry holystoned or scraped. This duty, as well as that of cleaning the ladders, hospitals and round houses shall be performed by a party who shall be taken in rotation from the adult males above 14, in the proportion of five to every 100 emigrants, and shall be considered as sweepers for the day. But the single women shall perform this duty in their own compartment, where a separate compartment is allotted to them, and the occupant of each berth shall see that his own berth is well brushed out.

7. Dinner shall commence at one o'clock p.m., and supper at six p.m.

8. The fires shall be extinguished at 7 p.m., unless otherwise directed by the master, or required for the use of the sick; and the emigrants shall be in their berths at 10 0 clock p.m., except under the permission or authority of the surgeon, or if there be no surgeon, of the master.

9. Three safety lamps shall be lit at dusk, and kept burning till 10 o'clock p.m.; after which hour two of the lamps may be extinguished, one being nevertheless kept burning at the main hatchway all night.

10. No naked light shall be allowed at any time, or on any account.

11. The scuttles and stern ports, if any, shall, weather permitting, be open at 7 o'clock a.m., and kept open till 10 p.m.; and the hatches shall be kept open whenever the weather permits.

12. The coppers and cooking utensils shall be cleaned every day.

13. The beds shall be well shaken and aired on deck at least twice a week.

14. The bottom boards of the berths, if not fixtures, shall be removed and dry-scrubbed, and taken on deck at least twice a week.

15. Two days in the week shall be appointed by the master as washing days; but no washing or drying of clothes shall on any account be permitted between decks.

16. On Sunday mornings, the passengers shall be mustered a. 10 0 clock a.m., and will be expected to appear in clean and decent apparel. The Lord's day shall be observed as religiously as circumstances will admit.

17. No spirits or gunpowder shall be taken on board by any passenger; and if either of those articles be discovered in the possession of a passenger, it shall be taken into the custody of the master during the voyage, and not returned to the passenger until he is on the point of disembarking.

18. No loose hay or straw shall be allowed below for any purpose.

19. No smoking shall be allowed between decks.

20. All gambling, fighting, riotous or quarrelsome behaviour, swearing, and violent language, shall be at once put a stop

to. Fire-arms, swords, and other offensive weapons shall, as soon as the passengers embark, be placed in the custody of the master.

21. No sailors shall be allowed to remain on the passenger deck among the passengers, except on duty.

22. No passenger shall go to the ships' cook-house without special permission from the master, nor remain in the forecastle among the sailors on any account.

23. In vessels not expressly required by the said "Passengers Act, 1852," to have on board such ventilating apparatus as therein mentioned, such other provision shall be made for ventilation as shall be required by the emigration officer at the port of embarkation, or in his absence, by the officers of customs.

24. And to prevent all doubts in the construction of this order m Council, it is hereby further ordered that the terms "United Kingdom," "passenger," "passenger ship," and "master", shall herein have the same significations as are assigned to them specifically in the said "Passenger Act, 1852". And the Right Hon. Sir John Pakington, Bart., one of her Majesty's Principal Secretaries of State, is to give the necessary directions herein accordingly. Wm. L. Bathurst.

APPENDIX 8

SURGEONS ON BOARD THE *GREAT BRITAIN*

Voy	Surgeon	Outward/ Inward	Voyage Dates	Itinerary
8	Andrew Alexander	Out & In	May 1852–Jun 1852	L–NY–L
9	Andrew Alexander	Out & In	Aug 1852–Apr 1853	L–M–S–M–L
10	Andrew Alexander	Out & In	Aug 1853–Feb 1854	L–M–S–M–L
11	Andrew Alexander	Out & In	Jun 1854–Jan 1855	L–M–L
12	William Gilmour	Out & In	Mar 1855–May 1856	Crimean War
13	Samuel Archer	Out & In	Feb 1857–Aug 1857	L–M–L
14	Samuel Archer	Out & In	Sep 1857–Apr 1858	Indian Mutiny
15	Charles Morice	Out & In	Jul 1858–Sep 1958	L–NY–L
16	Charles Morice	Out & In	Nov 1858–May 1859	L–M–L
17	Charles Morice	Out & In	Jul 1859–Aug 1859	L–NY–L
18	Charles Morice	Out & In	Dec 1859–May 1860	L–M–L
19	Charles Morice	Out & In	Jul 1860–Jan 1861	L–M–L
20	Thomas Hocken	Out & In	Feb 1861–Aug 1861	L–M–L
21	Thomas Hocken	Out	Oct 1861–Dec 1861	L–M–L
21	William Walsh	In	Feb 1862–Apr 1862	
22	Robert Newbold	Out & In	Jun 1862–Nov 1862	L–M–L
23	Andrew Alexander	Out & In	Jan 1863–Jul 1863	L–M–L

Voy	Surgeon	Outward/ Inward	Voyage Dates	Itinerary
24	Andrew Alexander	Out & In	Oct 1863–Apr 1864	L–M–L
25	Andrew Alexander	Out & In	May 1864–Oct 1864	L–M–L
26	Andrew Alexander	Out & In	Dec 1864–May 1865	L–M–L
27	Andrew Alexander	Out & In	Jul 1865–Dec 1865	L–M–L
28	Andrew Alexander	Out & In	Feb 1866–Jul 1866	L–M–L
29	Andrew Alexander	Out & In	Oct 1866–Mar 1867	L–M–L
30	Andrew Alexander	Out & In	May 1867–Oct 1867	L–M–L
31	Andrew Alexander	Out & In	Dec 1867–May 1868	L–M–L
32	Andrew Alexander	Out	Jul 1868–Sep 1868	L–M–L
	Alfred Puddicombe	In	Oct 1868–Dec 1868	
33	Alfred Puddicombe	Out & In	Feb 1869–Jul 1869	L–M–L
34	Alfred Puddicombe	Out & In	Aug 1869–Feb 1870	L–M–L
35	Alfred Puddicombe	Out	Mar 1870–May 1870	L–M–L
	Joshua Duke	In	Jun 1870–Aug 1870	
36	Robert McGowan	Out & In	Oct 1870–Mar 1871	L–M–L
37	William Smythe	Out & In	May 1871–Nov 1871	L–M–L
38	William Smythe	Out & In	Dec 1871–May 1872	L–M–L
39	William Smythe	Out & In	Jul 1872–Dec 1872	L–M–L
40	William Smythe	Out	Mar 1873–May 1873	L–M–L
	Alfred Puddicombe	In	Jul 1873–Sep 1873	
41	Alfred Puddicombe	Out	Oct 1873–Dec 1873	L–M–Br–L
	Edward Ascher	In	Feb 1874–Apr 1874	
42	Edward Ascher	Out & In	Jun 1874–Nov 1874	L–M–L
43	Edward Ascher	Out & In	Jan 1875–Jul 1875	L–M–L
44	Edward Ascher	Out & In	Aug 1875–Jan 1876	Lon–M–L

BI: Unpublished notes: Malcolm Lewis, 'Ship Surgeons and the *SS Great Britain*', November 2018

APPENDIX 9

A DESCRIPTION OF THE
GREAT BRITAIN IN 1870

Western Daily Press, Thursday 10 February 1870

ON BOARD THE GREAT BRITAIN. A correspondent, who has returned from Australia in the Bristol-built steamer *Great Britain*, has furnished us with the following interesting particulars: –

We feel pleasure in looking back at the career of valuable and useful lives in men, characters that leave their marks, for by them succeeding generations steer their courses or should do so, and help on the great work of the world's progress. As in men so in ships, and your ancient city added a monument to marine architecture when the great minds originated and carried out the idea of the construction of the *Great Britain*. This fine vessel is 3,500 tons burthen, was built in in 1843, and launched July 19th in that year. She first commanded by Lieut. Hoskin, R.N., who, unfortunately, ran the *Great Britain* on shore in Dundrum Bay, on the East Coast of Ireland; here buffetted by the waves and storms howling over, her faithful build withstood all; and, thanks to engineering skill, she was restored to her proper element. She was afterwards commanded by the late Capt. B. R. Mathews, a gentleman who filled most worthily the responsible position of Lloyd's agent to the port of Melbourne for ten years. By him she was piloted to America and Australia, each time carrying valuable lives and freight.

And now she acknowledges as her controlling spirit Lieut. John Gray, R.N.R., who has commanded her for 23 voyages to Australia, besides taking troops to India, 18 months in the Crimea, and a trip or two to America. During his command, Capt. Gray has carried upwards of 60,000 passengers – the population of a fair city – and without disasters of any moment. The carrying capacity of the ship is large, and during the Crimean war she had on board 2,000 Turkish soldiers and 90 officers, all of whom were provided with meals three times day. The *Great Britain*'s engines are by the famous makers, Penn, of London, 500 h.p. nominal, working to 1,200 horse, and are the engines that won the prize as marine engines at the World's fair. The cost of fitting the engine-room was £15,000. Her chief engineer, Mr Alexander Maiden, who is a native of Cheshire, has been in the vessel ten years, and leaves her this voyage to fill an important and lucrative appointment in Melbourne. Her chief officer, Mr Lamerson, like the captain, is a native of Shetland, and has filled the present office one voyage. This gentleman was an officer on the *Great Eastern* during one or more eventful voyages that ship made. The second officer, Mr Christer, is a native of Lancashire, and has been in the vessel three years. The third officer, Mr Congdon, is a native of Cornwall; and the fourth, Mr Dusantry, from Weston- super-Mare. All these officers are married men.

The *Great Britain* carries four engineers besides her chief, the second of whom, Mr Walker, will most probably succeed Mr Maiden.

Of the executive staff, Dr Puddicombe fills the medical department; and when we consider the *Great Britain* has on board at times 800 souls, "and the ills that flesh is heir to," the post of surgeon is no sinecure. Dr Puddicombe is a native of Moreton Hampstead, airy, health-restoring township in the centre of the most beautiful western county.

Now we come to the purser, a position nearly as responsible as the captain's. In his office the dietary portion of the ship is controlled, and whether the well-supplied tables in the saloon, or the more modest fare of the intermediate, all bear evidence of a master hand. Mr Unsworth, the gentleman alluded to, has been

17 years in the ship, and has as good an opinion of her now as the first day he joined. In his special duties he is ably seconded by John Campbell, the head steward, and 44 assistant stewards. A word of praise for this department: to the credit of all be it spoken that not one instance of insobriety has occurred since the *Great Britain* left Liverpool, and temptation to indulge is by no means small, for the "racks" are lined with the well-filled bottles of the passengers, which at all times the stewards and assistants have access. Mrs Lynch, the stewardess, has been in the vessel several years, and is an institution.

Then come to a department which all Englishmen like to have properly appointed. The *Great Britain*'s kitchen is a model one on perfect scale. Her chef de cuisine is a man of weight and excellence, and his cooking the theme of general admiration. No matter what the weather is, dinner is ready to the minute. Commend me, say I, to such an artiste, for such help us to enjoy life by turning to good uses the things the gods have given us.

To work this large vessel there are 143 names on the ship's articles, with the usual complement of quartermasters, etc,. A name does appear, however, that deserves more than a passing notice, and this is that of Joe Rodgers. Joe is a Maltese, and was on board the *Royal Charter* (s.s.) when that unfortunate ship was wrecked on the Welsh coast. Joe swam with a line through raging surf, and was instrumental under Providence in saving 29 lives. Since then he has "belonged" to this ship, and occupies the post of lamptrimmer, but he makes more money than the chief officer; he is washerman, and his services are in general request as such. In many other ways he is permitted perquisites, and everyone wishes him success in all he undertakes.

The *Great Britain* is ship-rigged, with double topsail yards; her main and fore yards square 104 ft; the diameter of her mainmast is 3ft 9in, and weighs about 16 tons. The expenses of this fine ship are £5,000 per month when she has her full complement of passengers. She undergoes a thorough overhaul every voyage, but such is the character of the work done by the late Mr Patterson, her builder, that her wear and tear are reduced to a minimum; her iron plates have wasted only the 32nd part of an inch in 26 years and only two have had to removed.

Every voyage the *Great Britain* is docked, and a thorough overhaul, to fit her for the next and succeeding trips, to claim the certificate necessary from the Steam Navigation Board, and to place her favourably before the eyes of shippers and insurance agents. The examination is most rigid, and, when finished, any vessel equal to the test is pronounced fit to withstand the usual wear and tear expected in the work any vessel has to perform.

In the colonies forming Australia the *Great Britain's* name is a household word. She carries passengers from Western Australia, South Australia, New South Wales, Tasmania, Queensland, and New Zealand, but in Victoria she is best known, as Hobson's Bay is her headquarters during her stay there. And now we come to the present voyage of the *Great Britain*. After having been detained 36 hours by tempestuous weather, Captain Gray took his anchor early on Friday morning, the 26th November, 1869, and steamed down the bay. The tide rip turned its curling wave, but yet the order was given, and after rolling and tumbling, she got through the boiling surf and discharged our pilot, Mr Rich, who was taken by the pilot boat *Rip*. In the offing the flying squadron, consisting of six ships, was waiting to enter. We stood on our course through the straits, and before daylight were abreast of Swan Island, when Captain Gray thought it prudent to lie to. Our course was then shaped for our present voyage. The Horn was passed on the 22nd of December, and we got thus far without seeing ice. On the morning of the 27th however, and when all thoughts of such disagreeable neighbours were far off, the look-out man reported icebergs, and 57 were counted from the fore-yard visible at one time. They were all to the eastward of us, and with the help of steam and sail we soon left them behind. The voyage was one of an unusually calm character. The Horn weather was of the most pacific character, and reminded the writer of the genial climate of the South Sea Islands rather than the dismal Tierra del Fuego.

The comforts of first-rate hotel have been almost surpassed on board the *Great Britain*. Capital beds, sweet smelling rooms, and thorough cleanliness throughout the ship are her marked characteristics; and the ease with which all is done is a marvel to the uninitiated. We have been a good deal under steam, and where strong winds were expected we found calms, the trade winds having been unusually far north, so that contrary expectations have been realised in this particular. A word to what this vessel carried as cargo. She had 1,900 bales of wool, 24,0000 ozs. of gold, 210 casks colonial butter (a novel export), 40 casks preserved meat, and 172 passengers of all classes. This small number returning is proof of the prosperity of the colony. Sailors sometimes say there are long passages when many gentlemen in sober black are on board, but this voyage there was not one, and instead we had five captains, who either had been wrecked or disposed of their ships. Among the other passengers we find agents for the Tasmanian and Queensland Governments, seeking money and labour; a South Pacific delegate, advocating the interests of the Fiji Islands; buyers of sugar machinery; a famous comedian, returning to England and the States; men who have made comfortable independence in mining and trading pursuits, and others less fortunate.

The connection of Captain Gray with the steamship *Great Britain* is a theme in itself. The knowledge of character, tact, presence of mind, never-wearying anxiety for ship and passengers, go far to make the *beau ideal* of a commander. Just fancy the responsibility of taking charge of 14 spirited "engaged young ladies" on a single voyage, and similar ones has this commander on each trip to the colonies. Gales of wind are but child's play compared with this.

We trust that Captain Gray will live long to fill his present post; ship and captain are perfectly in accord, and it will be a sad day when either separate. Thousands whom he has carried will re-echo this wish, for Captain Gray has done much to advance the interests of Australia as many who have large stakes in that favoured country.

The *Great Britain* arrived Liverpool on Friday morning last, the 4th instant, without having lost a rope-yarn, while other vessels were dismasted in the gale of yesterday fortnight. She leaves Liverpool again the end of March, and has a considerable number of passengers engaged for the-return voyage.

APPENDIX 10

TRADE AND WAGES IN VICTORIA, STATISTICS OF AUSTRALIA IN 1870

Western Daily Press, Friday 11 February 1870

TRADE AND WAGES VICTORIA.

A gentleman who has just returned from Australia on board the *Great Britain* has supplied us with the following particulars relating to the colony of Victoria. The information will be useful to emigrants: –

Of the colony of Victoria the present is, perhaps, a proper time to give a few statistics, it being, also, the queen over any of the others. The population of Victoria the 31st December, 1863, was 684,316, whom 384,859 were males and 299,457 females. There were supposed then to have been 1,908 aborigines only left. The squatters or graziers in 1865 were 1125, holding acres, for which they pay 2s yearly for each head, of cattle or horses, and eight pence for each sheep. The livestock in the colony in 1866 consisted of 121,051 horses; 621,337 cattle; 8,835,380 sheep; and 75,869 pigs. The land occupied, enclosed, cultivated, in 1866. consisted of holdings, 20,363; area in occupation, 6,785,225 acres; area enclosed, 5,357,962; area cultivated, 530,196. Land under principal crops – wheat, 178,628; oats, 102,817; barley, 6887; potatoes, 31,644; hay, 97,902; green crops, 55,830 acres. Quantity of produce raised, wheat, 3514,227; oats, 2,279,468;

barley, 153,490; potatoes, 83,166; hay, 96,101. Besides these were grown maize, rye, and bere, peas, beans, millet, soryham, turnips, mangold wurzel, beet, carrots, parsnips, cabbage, and onions. There are 4,057 acres of vines, which yield 176,959 gallons of wine; besides the vineyards, 6,654 acres, were in 1866 cultivated as gardens, and 3,449 acres as orchards. Of manufactories, Victoria, there are 80 breweries, 118 flour mills, and other works, 705. The breweries employ 405 horses, 228 drays and waggons; they used 4,797,375 lbs sugar, 469,131 bushels of malt, 631,391 lbs of hops, and they manufactured 7,988,978 gallons. Of mining machinery value was £1,773,271; manufactures, £1,341,026; agricultural machinery, £729,281. The amount of gold exported since the discovery of the gold fields, 1851-52, was 35,286,040 ozs. representing at £4 per oz value of £141,144,160. Wool, one of our largest staples, is thus represented: – From 1839 to 1865 it amounted to 449,695,704 lbs, valued at £31,366,652, besides tallow and hides, which reached £110 000. There are soap works and tanneries in the colony, which use large quantities of the raw material. There are 11 banks of issue in Victoria, three of which are local institutions, the rest branches from Great Britain. The deposits in the savings bank in 1865 amounted to £719,100. There are 13 insurance offices (marine and fire), and 20 building societies. The imports in 1865 were slightly above £13,000,000 value; exports in the same year, £13,150,748. There are 276 miles of railways (254 government, 22 private); their cost per mile was £35,689 14s. In Victoria the revenue and expenditure are £3,000,000 each – rather an expensive item to the public.

WAGES – Agricultural labourers get from 12s to 18s; ploughmen, 18s to 20s; reapers, per acre, with rations, 10s to 12s; mowers, 3s to 4s; thrashers, per bushel, with rations, to 1s to 5s.
PASTORAL LABOUR.
Shepherds per annum, with rations £30 to 35
Slock keepers, ditto 35 to 45
Hut keepers, ditto 25 to 30
Generally useful men on stations, per week, with rations 14s to 18s
Sheep washers, per week, with rations 15s

Shearers, per 100 sheep shorn, with rations 13s to 14s
ARTISAN LABOUR.
Masons per day, without rations 8s to 10s
Plasterers, ditto 8 0 10s
Bricklayers, ditto 8s 0 10s
Carpenters, ditto 7s to 9s
Blacksmiths, ditto 8s to 10s
Servants.
Males, and married couples without family per annum, with board and lodging £50 to 60
Married couples with family, per annum, with board and lodging £40 to 50
Men cooks on farms and stations, per week, with board and lodging 15s to £1 5s
Grooms, per annum, with board and lodging. £40 to 50
Gardeners, ditto £40 to 50
MISCELLANEOUS Labour
General labourers, per day, without rations. 5s to 7s
Stone-breakers, per cubic yard 3s to 4s 6d
Seamen, per month £4 to £6
(The latter price paid on this ship.)

The rent of a cottage, suitable for labouring man and his family, ranges in Melbourne from 3s to 7s per week. In the country the rate is somewhat lower; and on the gold fields a canvas tent is set up, or cottage of logs or split timber, with a bark roof, is erected upon Crown lands, so that rent is altogether saved.

Appendix 11

THE LAST SALE OF ITEMS FROM THE *GREAT BRITAIN*

Liverpool Mercury, Friday 10 February 1882

GREAT BRITAIN STEAMSHIP. WITHOUT RESERVE.
MONDAY, 20 th FEBRUARY.
The whole of the valuable stores and Furnishing Requisites of the famous Steamship *Great Britain* including plated Goods, Table Cutlery, Glass and China, Cooking Utensils, Mahogany Dining Tables. Settees, 80 Mirrors, Bookcases; Sideboards, Armchairs, Dinner, Tea and Desert Services, Toilet Ware. 1000 Sheets, 1000 Blankets, 130 Hair Mattresses. 300 Table Covers, 1000 Table Napkins. 2500 Towels, 250 Feather and Hair Pillows. 600 pillow Cases, 140 Carpets, 160 sets of curtains, Soup Tureens, 2000 plated Table forks and Spoons, 30 Plated Dish Covers, Entree Dishes, Tea and Coffee Pots, 2000 Dinner knives, 20000 pieces of Glass, 2500 pieces Dinner Ware. 500 volumes of Books.

WALKER ACKERLEY, & CO. will Sell by Auction, on Monday, the 30th instant, at eleven o'clock, ON BOARD THE GREAT BRITAIN STEAMSHIP, now lying at the HEAD OF THE GREAT FLOAT, BIRKENHEAD. The whole of the valuable STORES and FURNISHING REQUISITES, comprising upwards of 1500 PIECES OF ELECTRO-PLATE,

Including dish covers, entree dishes, soup tureens, cruet frames, tea and coffee services, toddy kettles, toast racks, fish carvers, table and dessert forks and spoons, napkin rings
1000 PIECES OF TABE CUTLERY,
2000 PIECES OF GLASS,
including decanters, wine glasses, tumblers, water bottles, etc
THE GLASS AND CHINA
Comprise decanters, wine glasses, hock classes, champagne glasses, water bottles, dinner ware, dessert were, tea ware, and toilet ware.
FURNITURE, etc.
Six mahogany dining tables, twelve mahogany settees, 80 mirrors, four Mahogany bookcases, arm chairs, sideboards, bracket tables, 140 carpets, 500 vols books etc
TABLE AND BED LINEN
3000 towels, 1000 sheets, 400 quilts. 1000 table napkins, 1000 blankets, 300 tablecloths, 700 bolster and pillow cases, 150 feather pillows, 130 hair pillows, 40 plush cushions, 120 hair mattresses

ENDNOTES

1 'Your Other Ship': A Second Vessel

1. Graeme J Milne, *Trade and Traders in Mid-Victorian Liverpool: Mercantile Business and the Making of a World Port* (Liverpool: Liverpool University Press, 2000), p. 23
2. Milne, *Trade and Traders in Mid-Victorian Liverpool*, p. 23
3. Ewan Corlett, *The Iron Ship: The Story of Brunel's SS Great Britain*, 2012 edn (Bristol: ss Great Britain, 1975); Andrew Kelly and Melanie Kelly, *Brunel, in Love with the Impossible: A Celebration of the Life, Work, and Legacy of Isambard Kingdom Brunel* (Bristol: Bristol Cultural Development Partnership, 2006); Denis Griffiths, Andrew Lambert, and Fred Walker, *Brunel's Ships* (London: Chatham Publishing, 1999).
4. Milne, *Trade and Traders in Mid-Victorian Liverpool*, p. 23
5. Isambard Brunel, *The Life of Isambard Kingdom Brunel Civil Engineer*, 2006 edn (Stroud: Nonsuch Publishing, 1870), p. 185
6. Brunel Institute, SS Great Britain Trust (BI): Great Western Steam Ship Company AGM 1839
7. Griffiths, Lambert, and Walker, *Brunel's Ships*, pp.64, 67
8. Christopher Claxton, *History and Description of the Steam Ship Great Britain* (Bristol, 1845).
9. Corlett, *The Iron Ship*, pp. 33, 35
10. Helen Doe, *The First Atlantic Liner: Brunel's Great Western Steamship.* (Stroud: Amberley Publishing 2017), p. 143

11. Brunel,*Life of Brunel*, p. 185
12. BI: Brunel Letterbook 1839
13. Brunel,*Life of Brunel*, p. 186
14. Sarah Palmer, 'Experience, Experiment and Economics: Factors in the Construction of Early Merchant Steamers', *Proceedings of the Atlantic Canada Shipping Project* (1977), pp. 237-38
15. Palmer, 'Experience, Experiment and Economics', p. 239
16. Palmer, 'Experience, Experiment and Economics', p. 238
17. BI: Great Western Steam Ship Company AGM 1840
18. Grahame Farr, *Shipbuilding in the Port of Bristol* (Greenwich, London: National Maritime Museum 1977)
19. Brunel,*Life of Brune*, p. 187
20. Brunel,*Life of Brunel*, p. 189
21. BI: Great Western Steam Ship Company AGM 1840
22. Corlett, *The Iron Ship*, p.121-22
23. BI: Great Western Steam Ship Company AGM 1840
24. BI: Great Western Steam Ship Company AGM 1840
25. BI: Great Western Steam Ship Company AGM 1840
26. BI: Great Western Steam Ship Company AGM 1840
27. BI: Great Western Steam Ship Company AGM 1840
28. BI: Great Western Steam Ship Company AGM 1840
29. BI: Great Western Steam Ship Company AGM 1840
30. James Taylor, *Creating Capitalism: Joint Stock Enterprise in British Politics and Culture, 1800-1870* (London: Royal Historical Society, 2006), p. 4
31. BI: Great Western Steam Ship Company AGM 1840
32. Corlett, *The Iron Ship*, p. 67
33. Griffiths, Lambert, and Walker, *Brunel's Ships*, p. 33.
34. Corlett, *The Iron Ship*, p. 71
35. Brunel, *Life of Brunel*, p. 190
36. R. Angus Buchanan, *Brunel: The Life and Times of Isambard Kingdom Brunel* (London: Hambledon and London, 2002), pp. 177-79
37. Andrew Lambert, 'Brunel, the Navy and the Screw Propeller' in Griffiths, Lambert, and Walker, *Brunel's Ships*
38. Corlett, *The Iron Ship*, p. 73
39. Corlett, *The Iron Ship*, p. 77
40. Buchanan, *Brunel*, p. 60

41. Corlett, *The Iron Ship*, p.123
42. BI: Great Western Steam Ship Company AGM 1841
43. BI: Great Western Steam Ship Company AGM 1841
44. BI: Great Western Steam Ship Company AGM 1841
45. BI: Great Western Steam Ship Company AGM 1841
46. Claxton, *History and Description*, p. 23

2 'This Extraordinary Vessel, the Wonder of the World

1. BI: Great Western Steam Ship Company AGM 1842
2. BI: Great Western Steam Ship Company AGM 1842
3. BI: Great Western Steam Ship Company AGM 1842
4. Ewan Corlett, *The Iron Ship: The Story of Brunel's SS Great Britain*, 2012 edn (Bristol: ss Great Britain, 1975), pp. 124-5
5. Helen Doe, *The First Atlantic Liner: Brunel's Great Western Steamship*. (Stroud: Amberley Publishing 2017), pp. 25-26
6. Corlett, *The Iron Ship*, pp. 124-25
7. BI: Great Western Steam Ship Company AGM 1843
8. Doe, *The First Atlantic Liner*, pp. 159-160
9. BI: Great Western Steam Ship Company AGM 1843
10. BI: Great Western Steam Ship Company AGM 1843
11. BI: Great Western Steam Ship Company AGM 1843
12. BI: Great Western Steam Ship Company AGM 1843
13. Andrew Lambert, 'Brunel, the Navy and the Screw Propeller', in Denis Griffiths, Andrew Lambert, and Fred Walker, *Brunel's Ships* (London: Chatham Publishing, 1999).
14. Lambert, 'Brunel, the Navy and the Screw Propeller', pp. 28-29
15. Lambert, 'Brunel, the Navy and the Screw Propeller', p.29
16. Lambert, 'Brunel, the Navy and the Screw Propeller', pp. 30-31
17. Lambert, 'Brunel, the Navy and the Screw Propeller', p. 27
18. T. R. Guppy Report, *Minutes and Proceedings of the Institution of Civil Engineers*, 4 March 1845, p. 157
19. Lambert, 'Brunel, the Navy and the Screw Propeller', pp. 41-43
20. BI: Great Western Steam Ship Company AGM 1844
21. Lambert, 'Brunel, the Navy and the Screw Propeller', p 27

22. R. Angus Buchanan, *Brunel: The Life and Times of Isambard Kingdom Brunel* (London: Hambledon and London, 2002), p. 178

23. BI: Great Western Steam Ship Company AGM 1844

24. *Bristol Times and Mirror* Saturday 22 July 1843

25. *Bristol Times and Mirror* Saturday 22 July 1843; *Bristol Journal* description quoted by Corlett, *The Iron Ship*, p.127

26. *Bristol Times and Mirror* Saturday 22 July 1843

27. *Bristol Times and Mirror* Saturday 22 July 1843

28. Corlett, *The Iron Ship*, p. 126

29. *Liverpool Mail* Saturday 6 January 1844

30. *Salisbury & Winchester Journal* 6 January 1844

31. BI: Great Western Steam Ship Company AGM 1844

32. *Hampshire Advertiser*, Saturday 7 December 1844

33. Isambard Brunel, *The Life of Isambard Kingdom Brunel Civil Engineer*, 2006 edn (Stroud: Nonsuch Publishing, 1870), p. 194

34. Buchanan, *Brunel*, p. 60

35. Christopher Claxton, *History and Description of the Steam Ship Great Britain* (Bristol, 1845)

36. Claxton, *History and Description*, pp. 3-5

37. Claxton, *History and Description*, pp. 89-90

38. Claxton, *History and Description*, pp. 89-90

39. Corlett, *The Iron Ship*, p. 111

40. Claxton, *History and Description*, pp. 3-5

41. Claxton, *History and Description*, pp. 89-90

42. Claxton, *History and Description*, pp. 3-5

43. Corlett, *The Iron Ship*, p. 113

44. Claxton, *History and Description*, p. 14

45. Corlett, *The Iron Ship*, p. 99

46. Corlett, *The Iron Ship*, p. 104

47. Claxton, *History and Description*.

48. Corlett, *The Iron Ship*, p. 101

49. V & A archive of Art & Design: Crace family, interior decorators: papers, 1691–1992

50. Claxton, *History and Description*, p. 14

51. Corlett, *The Iron Ship*, p. 103

52. Claxton, *History and Description*, p. 14

53. Claxton, *History and Description*, p. 3-5
54. Claxton, *History and Description*, pp. 89-90
55. Claxton, *History and Description*, pp. 89-90
56. T. R. Guppy Report, *Minutes and Proceedings of the Institution of Civil Engineers*, 4 March 1845, p. 161
57. Corlett, *The Iron Ship*, p. 136
58. *Hampshire Advertiser* Saturday 1 February 1845
59. T. R. Guppy Report, *Minutes and Proceedings of the Institution of Civil Engineers*, 4 March 1845, p. 164
60. T. R. Guppy Report, *Minutes and Proceedings of the Institution of Civil Engineers*, 4 March 1845, p. 164
61. *Hampshire Advertiser* Saturday 1 February 1845
62. Corlett, *The Iron Ship*, p.138
63. *Hampshire Advertiser* Saturday 1 February 1845
64. *Kentish Mercury* Saturday 3 May 1845
65. *Freeman's Journal* Monday 1 September 1845
66. *Perthshire Advertiser* Thursday 1 May 1845
67. Corlett, *The Iron Ship*, p. 143
68. *Perthshire Advertiser* Thursday 1 May 1845
69. Corlett, *The Iron Ship*, p. 143
70. *Perthshire Advertiser* Thursday 1 May 1845
71. *Kentish Mercury* Saturday 3 May 1845
72. Corlett, *The Iron Ship*, p. 144

3 'Monarch of the Ocean': The Great Britain Goes to Sea

1. *Hampshire Advertiser* 24 May 1845
2. *Royal Cornwall Gazette* Friday 23 May 1845
3. Ewan Corlett, *The Iron Ship: The Story of Brunel's SS Great Britain*, 2012 edn (Bristol: ss Great Britain, 1975), p.145-46
4. *Liverpool Mail* 5 July 1845. The newspaper referred to the engineer as 'Mr Guffee'
5. *Liverpool Mail* 5 July 1845
6. *Liverpool Mail* 5 July 1845
7. *Liverpool Mail* 5 July 1845
8. Corlett, *The Iron Ship*, p. 145-46
9. *Liverpool Mercury* Friday 11 July 1845
10. Corlett, *The Iron Ship*, p. 145-46
11. *Liverpool Mail* Saturday 26 July 1845

12. Corlett, *The Iron Ship*, p.119
13. Crosbie Smith, *Coal, Steam and Ships: Engineering, Enterprise and Empire on the Nineteenth Century Seas* (Cambridge: Cambridge University Press, 2018), p. 115
14. *New York Evening Post* 2 June 1845
15. Helen Doe, *The First Atlantic Liner: Brunel's Great Western Steamship.* (Stroud: Amberley Publishing 2017), pp. 79-84; James Hosken, *Autobiographical Sketch of the Public Career of Admiral James Hosken* (Penzance: Rodda, 1889)
16. Corlett, *The Iron Ship*, p. 146
17. BI: Database of passengers on Great Britain New York passenger list
18. Philippe Suchard (1827), *Mein Besuch Amerikas im Sommer 1824 (My visit to America in the summer of 1824)* (Switzerland: Aarau; Voegtli, M. 2003)
19. *Evening Post* August 12, 1845; *Banner of Ulster* Tues 2 September 1845
20. *The New York Herald*, quoted in *Inverness Courier* Wednesday 3 September 1845
21. *The New York Herald*, quoted in *Inverness Courier* Wednesday 3 September 1845
22. *The New York Herald*, quoted in *Inverness Courier* Wednesday 3 September 1845
23. *Wiltshire Independent* Thursday 4 September 1845
24. *Wiltshire Independent* Thursday 4 September 1845
25. *Wiltshire Independent* Thursday 4 September 1845
26. *Wiltshire Independent* Thursday 4 September 1845
27. *The Tribune* 28 August 1845
28. *The Tribune* 28 August 1845
29. Doe, *The First Atlantic Liner,* pp. 52-54, 195
30. *Evening Post* August 30, 1845
31. *Evening Post* August 30, 1845
32. BI: Passenger and Crew Database
33. Corlett, *The Iron Ship*, pp. 148-9
34. T. R. Guppy Report, *Minutes and Proceedings of the Institution of Civil Engineers*, 4 March 1845, p. 166
35. Corlett, *The Iron Ship*, p. 149
36. BI: Passenger and Crew Database

37. Corlett, *The Iron Ship*, p 150
38. Corlett, *The Iron Ship*, p 151
39. Corlett, *The Iron Ship*, p. 143
40. *Perthshire Advertiser* Thursday 1 May 1845
41. Corlett, *The Iron Ship*, p. 152
42. BI: Log of the *Great Britain*
43. BI: Log of the *Great Britain*
44. Corlett, *The Iron Ship*, p. 154
45. Corlett, *The Iron Ship*, p 157.
46. *Gore's Liverpool General Advertiser* Thursday 12 February 1846 and 16 April 1846
47. Denis Griffiths, *Brunel's Great Western* (Wellingborough: Patrick Stephens, 1985), p. 109
48. Corlett, *The Iron Ship*, p. 157

4 *'A Useless Saucepan': Disaster on an Irish Shore*

1. *Morning Advertiser* Friday 25 September 1846
2. *Morning Advertiser* Friday 25 September 1846
3. *Globe* Friday 25 September 1846
4. *Globe* Friday 25 September 1846
5. *Morning Advertiser* Friday 25 September 1846
6. *Globe* Friday 25 September 1846
7. *London Daily News* Friday 25 September 1846
8. *London Daily News* Friday 25 September 1846
9. L T C Rolt, *Isambard Kingdom Brunel*, 1st edition 1957 (London: Penguin, 1976), p. 272
10. *Ulster General Advertiser, Herald of Business and General Information*, Saturday 26 September 1846
11. *Evening Mail* Monday 28 September 1846
12. *Evening Mail* Monday 28 September 1846
13. *Evening Mail* Monday 28 September 1846
14. *Manchester Courier and Lancashire General Advertiser* Wednesday 30 September 1846
15. https://www.senate.gov/artandhistory/history/common/generic/VP_William_R_King.htm
16. *Hampshire Advertiser* Saturday 3 October 1846
17. *Hampshire Advertiser* Saturday 3 October 1846
18. *The Times* 1 October 1846

19. *The Times* 1 October 1846
20. Isambard Brunel, *The Life of Isambard Kingdom Brunel Civil Engineer*, 2006 edn (Stroud: Nonsuch Publishing, 1870) p. 195
21. Alison Winter, '"Compasses All Awry": The Iron Ship and the Ambiguities of Cultural Authority in Victorian Britain', *Victorian Studies*, 38 (1994).
22. *Evening Mail* Monday 28 September 1846
23. *Hampshire Advertiser* Saturday 24 October 1846
24. *Hampshire Advertiser* Saturday 24 October 1846
25. Ewan Corlett, *The Iron Ship: The Story of Brunel's SS Great Britain*, 2012 edn (Bristol: ss Great Britain, 1975), p. 161 quoting the *Mechanics Magazine*.
26. Winter, 'Compasses All Awry', p. 77
27. BI: Captain Hosken's Report 13 October 1846 and extracts from the minutes of the Great Western Steam Ship Company 20 October 1846
28. BI: Captain Hosken's Report 13 October 1846 and extracts from the minutes of the Great Western Steam Ship Company 20 October 1846.
29. *Hampshire Advertiser* Saturday 3 October 1846
30. Rolt, *Isambard Kingdom Brunel*, p. 273
31. Doe, *First Atlantic Line*, p. 154
32. *Shipping and Mercantile Gazette* Wednesday 28 October 1846
33. *Bristol Times and Mirror* Saturday 21 November 1846 and Saturday 28 November 1846
34. R. Angus Buchanan, *Brunel: The Life and Times of Isambard Kingdom Brunel* (London: Hambledon and London, 2002), p. 232
35. Brunel, *Life of Brunel*, p. 196
36. Rolt, *Isambard Kingdom Brunel*, p. 275
37. Brunel, *Life of Brunel*, p. 196
38. Brunel, *Life of Brunel*, p. 201
39. Brunel, *Life of Brunel*, p. 202-203
40. Brunel, *Life of Brunel*, p. 204 -206.
41. Information from Joanna Thomas, Maritime Curator, SS Great Britain Trust
42. Brunel, *Life of Brunel*, pp. 204-206.
43. Brunel, *Life of Brunel*, pp. 202 203

44. *Vindicator* Wednesday 23 December 1846
45. BI: Brunel Letter Books 5 January 1847
46. Corlett, *The Iron Ship*, p.176.
47. Brunel, *Life of Brunel*, pp. 208-209
48. Corlett, *The Iron Ship*, pp. 172-5; Denis Griffiths, Andrew Lambert, and Fred Walker, *Brunel's Ships* (London: Chatham Publishing, 1999), p. 88
49. *Bristol Times and Mirror* Saturday 4 March 1848
50. *Bristol Times and Mirror* Saturday 4 March 1848
51. *Bristol Times and Mirror* Saturday 4 March 1848
52. See Appendix 3 for more information on liabilities and risk
53. National Archives (TNA): RAIL 1149/60 Great Western Steam Ship Company AGM 1 March 1849
54. TNA: Rail 1149/60
55. James Hosken, *Autobiographical Sketch of the Public Career of Admiral James Hosken* (Penzance: Rodda, 1889).
56. Hosken, *Career of Admiral James Hosken*, pp. 33-39
57. Hosken, *Career of Admiral James Hosken*.
58. *Bristol Mercury* Saturday 8 July 1882
59. Corlett, *The Iron Ship*, p. 30
60. *Cardiff and Merthyr Guardian* 10 April 1847; *Perry's Bankrupt Gazette* 11 November 1848
61. *Bristol Times and Mirror* Saturday, 15 September 1849
62. *Dorset County Chronicle* 18 November 1858 'Sale of policy insurance in the West of England insurance office on the life of Thomas Richard Guppy Esq for £8,000 on which several bonuses have been declared'
63. *Bristol Mercury* Saturday 8 July 1882
64. *Bristol Mercury* Saturday 8 July 1882
65. *John o' Groat Journal* Friday 22 March 1901

5 *'This Majestic Vessel, Queen of the Ocean': New Owners*

1. British Parliamentary Papers (BPP): 19th General Report of the Emigration Commissioners 1859
2. Graeme J Milne, *Trade and Traders in Mid-Victorian Liverpool: Mercantile Business and the Making of a World Port* (Liverpool: Liverpool University Press, 2000), p. 176

3. Milne, *Trade and Traders in Mid-Victorian Liverpool*, p. 58; F J A Broeze, 'The Cost of Distance: Shipping and the Early Australian Economy, 1788-1850', *Economic History Review*, 28 (1975), p. 592

4. Ewan Corlett, *The Iron Ship: The Story of Brunel's SS Great Britain*, 2012 edn (Bristol: ss Great Britain, 1975), p. 182

5. Corlett, *The Iron Ship*.

6. Milne, *Trade and Traders in Mid-Victorian Liverpool*, p. 59

7. Corlett, *The Iron Ship*, p. 188

8. Alison Winter, '"Compasses All Awry": The Iron Ship and the Ambiguities of Cultural Authority in Victorian Britain', *Victorian Studies*, 38 (1994), pp. 81-82

9. Winter, 'Compasses All Awry', p. 82

10. Winter, 'Compasses All Awry', p.73

11. Winter, 'Compasses All Awry', p. 75

12. Winter, 'Compasses All Awry', p. 75

13. Winter, 'Compasses All Awry', p. 83

14. *Morning Post* Wednesday 10 March 1852

15. Winter, 'Compasses All Awry', p. 90

16. *London Evening Standard* Wednesday 24 March 1852

17. *Morning Post* Friday 26 March 1852; *Manchester Courier and Lancashire General Advertiser* Saturday 27 March 1852

18. *Morning Post* Friday 26 March 1852; *Manchester Courier and Lancashire General Advertiser* Saturday 27 March 1852

19. *Belfast Mercury* Saturday 27 March 1852

20. *Shipping and Mercantile Gazette* Friday 15 July 1853

21. *Illustrated London News* 24 April 1852

22. *Illustrated London News* Saturday 29 May 1852

23. BI: Passenger and crew database and Global Stories

24. *Illustrated London News* Saturday 29 May 1852

25. *Cork Constitution* Tuesday 20 July 1852

26. Corlett, *The Iron Ship*, p. 196

27. Helen Doe, *The First Atlantic Liner: Brunel's Great Western Steamship*. (Stroud: Amberley Publishing 2017), pp. 84-85

28. *Illustrated London News* Saturday 21 February 1852

29. Milne, *Trade and Traders in Mid-Victorian Liverpool*, p. 190

30. BI: TBM Park memoir of 1852 voyage

31. BI: Edward Towle Diary
32. BI: Edward Towle Diary
33. BI: Olcher Fedden diary
34. BI: Edward Towle Diary
35. Corlett, *The Iron Ship*, p.198-199
36. M K Stammers, 'Letters from the Great Britain 1852', *The Mariner's Mirror*, 62 (1976).
37. BI: Reginald Bright diary p.133
38. Stammers, 'Letters from the Great Britain 1852'
39. Corlett, *Iron Ship*, p. 201
40. BI: Olcher Fedden diary, Saturday, 6 November 1852
41. BI: Olcher Fedden Tuesday 9 November 1852
42. Corlett, *The Iron Ship*, p. 201
43. *Bath Chronicle and Weekly Gazette* Thurs 7 April 1853
44. Corlett, *The Iron Ship*, p. 201
45. Corlett, *The Iron Ship*, p. 205
46. Robert Saddington diary Tuesday, 11 October 1853
47. *Illustrated Sydney News* Saturday 12 November 1853
48. *Western Times*, 23 July 1853
49. Doe, *The First Atlantic Liner*, p. 166
50. Corlett, *The Iron Ship*, pp.195-99; BI: Captains Box and funeral notice in *The Age* 27 April 1869
51. BI: Reginal Bright diary p.133
52. *Bucks Herald* Sat 3 June 1854
53. *Liverpool Daily Post*. Reported in *Grantham Journal* Sat 5 Nov 1854
54. *Liverpool Daily Post*. Reported in *Grantham Journal* Sat 5 Nov 1854
55. BI: DM 1306/5/2-3 correspondence between Brunel and Christopher Claxton
56. TNA: BT 41/36 /209 Australian Steam Navigation Company 20 Dec 1851
57. TNA: BT 41/36/209 Australian Steam Navigation Company
58. TNA: BT 41/36/209 Australian Steam Navigation Company
59. TNA: CO 201/458 applying for more runs from bimonthly to monthly service.
60. TNA: TS 45/58 Treasury solicitors file 4 November 1852. Delay to HM Mails carried in steam ship Melbourne, 1852

61. TNA: BT 107/106 London Shipping Registers
62. *Adelaide Times* 5 May, 10 May 1853; *Sydney Morning Herald* Mon 29 August 1853
63. L T C Rolt, *Isambard Kingdom Brunel*, 1st edition 1957 (London: Penguin, 1976), p. 308.
64. *The Times* Monday 9 August 1852
65. *The Times* Monday 9 August 1852
66. TNA: HO 107/2443 1851
67. *Monmouthshire Merlin* 9 Oct 1847 quoting *Railway Gazette*
68. BI: DM 1306/5/2-3 correspondence between Brunel and Christopher Claxton
69. *Liverpool Albion* quoted in *Lincolnshire Chronicle* 1 Jan 1854

6 Her Majesty's Troopship: War Service

1. Hew Strachan, 'Soldiers, Strategy and Sebastopol', *The Historical Journal*, 21 (1978), 303-25
2. Mike Stammers, *The Emigrant Clippers to Australia: The Black Ball Line, Its Operation, People and Ships, 1852-1871* (Barnoldswick: Milestone Research), p. 56
3. *The Times* 11 January 1855
4. Corlett, *The Iron Ship*, p. 207-8
5. Corlett, *The Iron Ship*, p. 208
6. *Illustrated London News* 13 January 1855
7. Alston Kennerley, 'Stoking the Boilers: Firemen and Trimmers in British Merchant Ships, 1850-1950', *International Journal of Maritime History*, XX (2008), p. 209
8. BI: Heywood Bright letter
9. Richard J. Evans, 'Epidemics and Revolutions: Cholera in Nineteenth-Century Europe', *Past & Present* (1988), pp. 127, 134; Matthew Smallman-Raynor and Andrew D. Cliff, 'The Geographical Spread of Cholera in the Crimean War: Epidemic Transmission in the Camp Systems of the British Army of the East, 1854–55', *Journal of Historical Geography,* 30 (2004), p. 42
10. Smallman-Raynor and Cliff, 'The Geographical Spread of Cholera in the Crimean War', p. 33
11. Sandra Hempel, *The Medical Detective: John Snow, Cholera, and the Mystery of the Broad Street Pump* (Granta Books, 2006)

12. BI: Mike Hinton, 'Cholera on *Great Britain* during July 1855'
13. BI: Mike Hinton, 'Cholera on *Great Britain* during July 1855'
14. BI: Heywood Bright letters
15. Ewan Corlett, *The Iron Ship: The Story of Brunel's SS Great Britain*, 2012 edn (Bristol: ss Great Britain, 1975), p. 207; BI: *Great Britain Times* published in 1865; Doe, *Brunel's Ships and Boats*
16. *Maidstone Journal and Kentish Advertiser* Monday 19 September 1864
17. Arvel B. Erickson and Fr. John R. McCarthy, 'The Yelverton Case: Civil Legislation and Marriage', *Victorian Studies,* Vol. 14, No. 3 (Mar., 1971), pp. 275, 289-290
18. *Illustrated London News* Saturday 9 March 1861
19. *Dundee, Perth, and Cupar Advertiser* Tuesday 20 January 1863
20. Arvel B. Erickson and Fr. John R. McCarthy, The Yelverton Case: Civil Legislation and Marriage, *Victorian Studies,* Vol. 14, No. 3 (Mar., 1971), pp. 275, 289-290
21. Albert Nicholson, revised by Catherine Pease-Watkin, 'Longworth, Maria Theresa (1833–1881)', *Oxford Dictionary of National Biography* (2004); Erickson and McCarthy, 'The Yelverton Case', pp. 275, 289-290
22. Corlett, *The Iron Ship,* pp. 213-214
23. *Lancaster Gazette* 21 Feb 1857
24. BI: Diary of Samuel Archer surgeon
25. *Evening Mail*, Friday 8 January 1858
26. BI: Diary of Samuel Archer surgeon.
27. *Liverpool Daily Post* Tuesday 13 July 1858
28. *Glasgow Morning Journal* Friday 15 October 1858
29. *New York Times* 11 July 1859
30. *New York Times* 20 July 1859
31. *New York Times* 20 July 1859
32. Grahame Farr, *Shipbuilding in the Port of Bristol* (Greenwich, London: National Maritime Museum 1977) pp 9-10
33. *Staffordshire Advertiser* 12 November 1864; *Bristol Times and Mirror* 1 February 1864; *Bristol Times and Mirror* 4 April 1868

7 *'This Splendid Craft': Owners, Masters and Crew*

1. TNA: BT 107/329 Registry of Ships and Seamen Liverpool
2. Milne, *Trade and Traders*
3. Doe, *The First Atlantic Liner*; BI: Peter Revell, 'Robert Bright Biographical Notes', p. 14
4. Doe, *The First Atlantic Liner*, pp. 140-41
5. Revell, 'Robert Bright Biographical Notes', p.24
6. Revell, 'Robert Bright Biographical Notes', p. 24
7. Revell, 'Robert Bright Biographical Notes', pp.38–39
8. BI: Great Western Steamship Company AGM papers
9. *Southern Reporter and Cork Commercial Courier* Tuesday 9 April 1850
10. *Nottingham Journal* Friday 21 June 1850
11. *Liverpool Mail* Saturday 24 January 1852
12. *Nottingham Journal* Friday 13 May 1853
13. Mike Stammers, *The Emigrant Clippers to Australia: The Black Ball Line, Its Operation, People and Ships, 1852-1871* (Barnoldswick: Milestone Research), pp. 24, 26, 34
14. London Metropolitan Archives (LMA): Antony Gibbs Collection, Gibbs, Bright
15. LMA: Gibbs, Bright
16. *Shipping and Mercantile Gazette* Wednesday 12 March 1856
17. *Glasgow Herald* Saturday 2 April 1859
18. Stammers, *The Emigrant Clippers to Australia*
19. *Glasgow Herald* Saturday 2 April 1859
20. BI: Passenger and crew database
21. M K Stammers, 'Letters from the Great Britain 1852', *The Mariner's Mirror*, 62 (1976)
22. Stammers, *Emigrant Clippers to Australia*, p. 40
23. *Western Daily Press* Thursday 10 February 1870
24. BI: Passenger and crew database
25. BI: Passenger and crew database
26. BI: Notes from Global Stories Project
27. *Manchester Times* Saturday 21 September 1844
28. *Globe* 20 October 1852
29. 1851 UK census

30. BI: Passenger and crew database and Census data
31. *Liverpool Daily Post* 16 April 1867
32. *Western Daily Press* Thursday 10 February 1870
33. Stammers, *Emigrant Clippers to Australia*
34. BI: John Campbell diary
35. BI: Notes from Global Stories Project Global stories
36. Robin Haines, *Doctors at Sea: Emigrant Voyages to Colonial Australia* (Basingstoke: Palgrave MacMillan, 2005), p. 77
37. Haines, *Doctors at Sea*, p. 81
38. Haines, *Doctors at Sea*, p. 81
39. BI: Malcolm Lewis, 'Ship Surgeons and the *ss Great Britain,*' November 2018, p. 92
40. Haines, *Doctors at Sea*, p. 84
41. BI: Malcolm Lewis, 'Ship Surgeons', p. 42
42. BI: John Campbell diary 1871
43. John Crossley, 'Death and Disease on the SS Great Britain', *Proceedings of Bristol Medico-Historical Society*, 3 (1999), p. 92
44. BI: John Campbell diary
45. Crossley, 'Death and Disease', pp, 89-92
46. Information from Joanna Thomas, Maritime curator, SS Great Britain Trust
47. BI: Malcolm Lewis, 'Ship Surgeons', p. 54
48. BI: Samuel Archer diary
49. BI: Malcolm Lewis, 'Ship Surgeons', p. 60
50. G M Strathern, 'One sketch of his life' quoted by Malcolm Lewis, 'Ship Surgeons'
51. BI: Malcolm Lewis, 'Ship Surgeons', p. 80
52. Stammers, *Emigrant Clippers to Australia*, p. 87
53. Helen Doe, 'Power, Authority and Communications: The Role of the Master and the Managing Owner in Nineteenth Century Merchant Shipping', *International Journal of Maritime History*, Vol XXV, No 1, June 2013
54. John Gray Masters certificate application
55. BI: Peter Revell 'John Gray Biographical Notes'
56. BI: W D Waters diary Saturday, 4 January 1868.
57. BI: Extract from 'Life's Panorama' by Mr J A Gurner
58. BI: Peter Revell 'John Gray Biographical Notes'

59. *Luton Times and Advertiser* Saturday 18 June 1870
60. *Leeds Intelligencer* Saturday 27 May 1865
61. BI: Diary of John Campbell
62. *Liverpool Albion* reported in *Ballymena Observer* Saturday 4 January 1873
63. *Manchester Evening News* Friday 27 December 1872
64. *Morning Post* Saturday 28 December 1872
65. *Liverpool Mail* Saturday 22 February 1873
66. *Liverpool Mail* Saturday 22 February 1873
67. James Crompton (ed.), *A Journal of a Honeymoon Voyage in SS Great Britain* (SS Great Britain Project: 1992), p. 11

8 'Queen of the Waters': Passengers to Australia

1. Robin Haines & Ralph Shlomowitz (1992) Immigration from the United Kingdom to colonial Australia: A statistical analysis, *Journal of Australian Studies*, 16:34, 43-52, p 45
2. F J A Broeze, 'The Cost of Distance: Shipping and the Early Australian Economy, 1788-1850', *Economic History Review*, 28 (1975), p. 240
3. Stammers, *Emigrant Clippers to Australia*, p. 15
4. Broeze, 'The Cost of Distance', p. 240
5. BI: Database of Passengers and Crew
6. BI: Database of Passengers and Crew
7. Milne, *Trade and Traders*, p. 191
8. Milne, *Trade and Traders*, p. 191
9. Robert Scally, 'Liverpool Ships and Irish Emigrants in the Age of Sail', *Journal of Social History*, 17 (1983), p. 14
10. Milne, *Trade and Traders*, p. 191
11. Robin Haines, *Doctors at Sea: Emigrant Voyages to Colonial Australia* (Basingstoke: Palgrave MacMillan, 2005), p. 118
12. Haines, *Doctors at Sea*, p. 119
13. Haines, *Doctors at Sea*, p. 119
14. Haines, *Doctors at Sea*, p. 119
15. Haines, *Doctors at Sea*, p. 123
16. Haines, *Doctors at Sea*, p. 123
17. Haines, *Doctors at Sea*, p. 125
18. BI: Database of Passengers and Crew additional information supplied by descendent

19. Ancestry.com. 'Victoria, Australia, Assisted and Unassisted Passenger Lists, 1839-1923'

20. 1851 census Sampford Peverell Dorset; 1861 census Sampford Peverell Dorset

21. 1861 census Widnes St Mary

22. *Liverpool Mercury* 22 November 1861

23. BPP Return of Immigration Officers and Medical Inspectors at Ports in the United Kingdom dated 15 February 1854

24. Kate Mathew, 'The Female Middle Class Emigration Society: Governesses in Australia: A Failed Vision?', *Journal of Australian Colonial History*, 14 (2012), pp 107-130

25. Matthew, 'Governesses in Australia', p. 110

26. Mathew, 'Governesses in Australia', p. 112

27. Mathew, 'Governesses in Australia', p. 118

28. BI: Diary of Rachel Henning

29. Mathew, 'Governesses in Australia', p. 113

30. Mathew, 'Governesses in Australia', p. 124

31. M K Stammers, 'Letters from the Great Britain 1852', *The Mariner's Mirror*, 62 (1976).

32. Haines, *Doctors at Sea*, p. 127

33. Haines, *Doctors at Sea*, p. 129

34. BI: Diary of Rosamond D'Ouseley Tuesday 5 September 1869

35. *Liverpool Mercury* 23 May 1865

36. BI: Diary of Rachel Henning

37. Janet C Myers, 'Performing the Voyage Out: Victorian Female Emigration and the Class Dynamics of Displacement', *Victorian Literature and Culture*, 29 (2001), p. 131

38. Email from James Holmes a Court, 18 January 2019

39. Claire Connor, 'Passengers, Emigrants and Modern Men: A Social History of the 1852 Voyage of SS *Great Britain* from Liverpool to Melbourne' PhD, University of Bristol, 2014, p. 346

40. Information on the Veness family kindly supplied by Jennie Manners in email dated 29 May 2018

41. BI: Diary of Allan Gilmour

42. BI: Diaries of Charles Chomley and Louise Buchan

43. BI: Diary of Daniel Higson

44. Lesley Trotter, *The Married Widows of Cornwall: The Story of the Wives Left behind by Emigration* (Humble History Press, 2018)

45. BI: *The Cabinet*: 'a repository of facts, figures, and fancies relating to the voyage of the 'Great Britain" S.S. from Liverpool to Melbourne with The Eleven of All England and other distinguished passengers', 1861

46. Philip Payton, *Making Moonta: The Invention of Australia's Little Cornwall* (University of Exeter, Exeter, 2007).

47. BPP: Return of Vessels and Number of Immigrants for South Australia and Victoria 1847 to 1852

48. BI: Information from Global Stories Project

49. Andy Collier, *Across the Oceans with the English Cricketers to Australia and New Zealand 1861 to 1962* (2018)

50. Collier, *Across the oceans with the English Cricketers*, p.14

51. www.ssgreatbritain.org

52. Collier, *Across the oceans with the English Cricketers*, p. 16

53. BI: Diary of Edward Grace

54. *John o' Groat Journal* Thursday 25 April 1878

55. *The Albatross: Record of voyage of the Great Britain Steam Ship from Victoria to England in 1862* (Stirling: Charles Rogers & Co, 1863)

56. BI: Diary of Rosamond D'Ouseley Sunday 15 August 1869

57. *Empire* (Sydney) 17 March 1853

58. Wrexham Guardian and Denbighshire and Flintshire Advertiser Saturday 1 June 1864

59. Crompton, James (ed.), *A Journal of a Honeymoon Voyage in SS Great Britain* (SS Great Britain Project:1992), p 6.

60. BI: Samuel Archer diary Wednesday 20 May 1857

61. James Crompton (ed.), *A Journal of a Honeymoon Voyage in SS Great Britain* (SS Great Britain Project: 1992), p 5.

62. Crompton, *A Journal of a Honeymoon Voyage*, p.9

63. Crompton, *A Journal of a Honeymoon Voyage*, p.9

64. Crompton, *A Journal of a Honeymoon Voyage*, p 11, 15

65. Crompton, *A Journal of a Honeymoon Voyage*, p 24

9 'This Celebrated Steam Ship': Finished with Engines

1. Mike Stammers, *The Emigrant Clippers to Australia: The Black Ball Line, Its Operation, People and Ships, 1852-1871* (Barnoldswick: Milestone Research) p. 76

2. *Leeds Intelligencer* Saturday 27 May 1865; BI: Liverpool Customs Bills of Entry

3. Stammers, *Emigrant Clippers to Australia*, pp. 85 to 87

4. Sari Maenpaa 'Combining Business and Pleasure? Cotton Brokers in the Liverpool Business Community in late 19th Century', in Jarvis, A and Robert Lee (eds) *Trade Migration and Urban Networks* (St Johns, 2008)

5. BI: Archer diary Wednesday 20 May 1857

6. Ancestry.com: Masters and Mates certificate

7. *Liverpool Mercury* Friday 5 June 1874

8. Crosbie Smith, *Coal, Steam and Ships: Engineering, Enterprise and Empire on the Nineteenth Century Seas* (Cambridge: Cambridge University Press, 2018), pp. 336-41

9. *Irish Times* Saturday 19 April 1873

10. *Liverpool Mercury* Thursday 3 February 1870.

11. *Irish Times* Saturday 19 April 1873

12. *Irish Times* Saturday 19 April 1873

13. *Irish Times* Saturday 19 April 1873

14. *Irish Times* Saturday 19 April 1873

15. *Irish Times* Saturday 19 April 1873

16. *The Times* 28 April 1873

17. *Irish Times* Saturday 19 April 1873

18. Stammers, *Emigrant Clippers to Australia*, pp. 65-66.

19. BI: Poster announcing *Great Britain* departure for Melbourne and Brisbane

20. *Liverpool Mercury* January 1875, Friday 4 June 1875

21. *Bradford Observer* 24 June 1875, *Liverpool Mercury* 22 June 1875

22. *Lloyd's List* Wednesday 24 February 1875

23. *Morning Post* Tuesday 6 March 1877; *Southern Reporter* Thursday 9 March 1876; *London Evening Standard* Saturday 22 April 1876; *Naval & Military Gazette and Weekly Chronicle of the United Service* Wednesday 18 November 1874

24. Stammers, *Emigrant Clippers to Australia*, pp 92-95

25. Stammers, *Emigrant Clippers to Australia*, pp 101-102

26. Stammers, *Emigrant Clippers to Australia*, p. 17

27. *Manchester Courier and Lancashire General Advertiser* Saturday 30 July 1881

28. *Southend Standard and Essex Weekly Advertiser* Friday 23 September 1881

29. *Liverpool Mercury* Friday 10 February 1882

30. Elizabeth Neill, *Fragile Fortunes; The Origins of a Great British Merchant Family*, Wellington: Halsgrove, 2008), p. 383

31. Corlett, *The Iron Ship*, 1975), p. 231-2

32. Ancestry.com: Morris Masters certificate; Information from Joanna Thomas, Maritime Curator, SS Great Britain Trust

33. Ancestry.com: Stap master certificate and census data

34. Corlett, *The Iron Ship*, p. 233

35. Corlett, *The Iron Ship*, p. 233; BI: Passenger and crew database

36. Corlett, *The Iron Ship*, pp. 234-35; BI: Liverpool Customs Bills of Entry

37. *Western Daily Press* Thursday 21 January 1886

38. Corlett, *The Iron Ship*, p. 235

39. James Muirhead, 'The ss Great Britain – an Object in Stasis: Space, Place & Materiality, 1886-2015', unpublished PhD Thesis, University of Bristol, p. 72

40. Muirhead, 'An Object in Stasis: Space, Place & Materiality', p. 72

41. Muirhead, 'An Object in Stasis: Space, Place & Materiality', quoted on p. 69

42. *The Times* 14 December 1886

43. *Bristol Mercury* 18 May 1887

44. *London Evening Standard* 18 January 1898; J. Ann Hone, 'Bright, Charles Edward (1829–1915)', *Australian Dictionary of Biography*, National Centre of Biography, Australian National University, http://adb.anu.edu.au/biography/bright-charles-edward-149/text4499, published first in hard copy 1969, accessed online 7 January 2019.

45. Muirhead, 'An Object in Stasis: Space, Place & Materiality', p. 77

10 *'A Unique Nautical Antique': In the Twentieth Century and Today*

1. TNA: BT 107/106 London Foreign Shipping Registers

2. *Bath Chronicle and Weekly Gazette* Thursday 1 June 1911

3. Ewan Corlett, *The Iron Ship: The Story of Brunel's SS Great Britain*, 2012 edn (Bristol: ss Great Britain, 1975) pp. 237-238

4. Muirhead, James, 'The *SS Great Britain* – an Object in Stasis: Space, Place & Materiality, 1886–2015', PhD Thesis, University of Bristol pp. 79- 80
5. Muirhead, 'An Object in Stasis: Space, Place & Materiality', p. 81
6. Muirhead, 'An Object in Stasis: Space, Place & Materiality', p. 83
7. Muirhead, 'An Object in Stasis: Space, Place & Materiality', p. 85
8. Muirhead, 'An Object in Stasis: Space, Place & Materiality', p. 85
9. Chris Young, *The Incredible Journey: The SS Great Britain Story 1970-2010* (Bristol: SS Great Britain Trust, 2010) pp. 15–16
10. Muirhead, 'An Object in Stasis: Space, Place & Materiality,' p. 70.
11. Muirhead, 'An Object in Stasis: Space, Place & Materiality', p. 90.
12. Young, *The Incredible Journey*, p. 16
13. Young, *The Incredible Journey*, p. 17
14. Muirhead, 'An Object in Stasis: Space, Place & Materiality', p. 96
15. Muirhead, 'An Object in Stasis: Space, Place & Materiality', p. 113.
16. Young, *The Incredible Journey*, p. 17
17. Muirhead, 'An Object in Stasis: Space, Place & Materiality', p. 369
18. Muirhead, 'An Object in Stasis: Space, Place & Materiality', pp. 107-8
19. Basil Greenhill, *The Merchant Schooners* (Harper Collins, 1988), First Edition 1951
20. Chris Young, *The Incredible Journey: The SS Great Britain Story 1970-2010* (Bristol: SS Great Britain Trust, 2010), p. 20
21. Young, *The Incredible Journey*, p. 21
22. Richard Goold-Adams, *The Return of the Great Britain* (London: Weidenfield and Nelson, 1976), pp. 12-13
23. Young, *The Incredible Journey*, p. 22
24. Goold-Adams, *The Return of the Great Britain*, p. 23
25. Goold-Adams, *The Return of the Great Britain*, pp. 25-26

26. Corlett, *The Iron Ship*, pp 243-4
27. Goold-Adams, *The Return of the Great Britain*, p 28
28. Goold-Adams, *The Return of the Great Britain*, pp. 31-32
29. Muirhead, 'An Object in Stasis: Space, Place & Materiality', p. 183
30. Muirhead, 'An Object in Stasis: Space, Place & Materiality', pp. 175-6
31. Young, *The Incredible Journey*, p. 24
32. Young, *The Incredible Journey*, p. 24
33. Goold-Adams, *The Return of the Great Britain*, p. 33
34. Goold-Adams, *The Return of the Great Britain*, p. 35
35. Goold-Adams, *The Return of the Great Britain*, p. 42
36. Goold-Adams, *The Return of the Great Britain*, p. 42
37. Goold-Adams, *The Return of the Great Britain*, pp.43-44.
38. Goold-Adams, *The Return of the Great Britain*, p. 39
39. *Birmingham Daily Post* Wednesday 21 June 1972
40. Young, *The Incredible Journey*, pp. 25-26
41. Goold-Adams, *The Return of the Great Britain*, p. 52-53
42. Goold-Adams, *The Return of the Great Britain*, p. 55
43. Goold-Adams, *The Return of the Great Britain*, p. 60
44. Goold-Adams, *The Return of the Great Britain*, p. 61
45. Muirhead, 'An Object in Stasis: Space, Place & Materiality', p. 80
46. Muirhead, 'An Object in Stasis: Space, Place & Materiality', p. 167
47. Information supplied by Joanna Thomas, Maritime Curator, SS Great Britain Trust
48. Goold-Adams, *The Return of the Great Britain*, pp. 66-67
49. Corlett, *The Iron Ship*, p. 248; Goold-Adams, *The Return of the Great Britain*, pp. 73-4
50. Goold-Adams, *The Return of the Great Britain*, p. 73
51. Goold-Adams, *The Return of the Great Britain*, p. 81; Corlett, *The Iron Ship*, p. 249
52. Goold-Adams, *The Return of the Great Britain*, p. 62
53. Muirhead, 'An Object in Stasis: Space, Place & Materiality', pp. 117-18
54. Muirhead, 'An Object in Stasis: Space, Place & Materiality', pp. 117-18

55. Muirhead, 'An Object in Stasis: Space, Place & Materiality', pp. 117-18
56. Muirhead, 'An Object in Stasis: Space, Place & Materiality', pp. 117-8
57. Muirhead, 'An Object in Stasis: Space, Place & Materiality', p. 121
58. Muirhead, 'An Object in Stasis: Space, Place & Materiality', p. 127
59. http://www.falklands-museum.com/fennia.html
60. Goold-Adams, *The Return of the Great Britain*, pp. 80-81
61. Goold-Adams, *The Return of the Great Britain*, p. 81
62. Muirhead, 'An Object in Stasis: Space, Place & Materiality', p. 117
63. Muirhead, 'An Object in Stasis: Space, Place & Materiality', pp. 132-33
64. Muirhead, 'An Object in Stasis: Space, Place & Materiality', p. 138
65. *The Times* 3 October 1969
66. Goold-Adams, *The Return of the Great Britain*, p. 46
67. Muirhead, 'An Object in Stasis: Space, Place & Materiality', p. 114
68. Goold-Adams, *The Return of the Great Britain*, p. 91
69. Muirhead, 'An Object in Stasis: Space, Place & Materiality', p 167
70. Muirhead, 'An Object in Stasis: Space, Place & Materiality', p. 170
71. Goold-Adams, *The Return of the Great Britain*, p. 100
72. Corlett, *The Iron Ship*, pp. 251-55
73. *The Times* 24 June 1970 article by Trevor Fisher Lock
74. *The Times* 24 June 1970 article by Trevor Fisher Lock
75. *The Times* 24 June 1970 article by Trevor Fisher Lock; Muirhead, 'An Object in Stasis: Space, Place & Materiality', p. 190
76. Muirhead, 'An Object in Stasis: Space, Place & Materiality', p. 193
77. *The Times* 24 June 1970 article by Trevor Fisher Lock; Muirhead, 'An Object in Stasis: Space, Place & Materiality', p. 190

78. Corlett, *The Iron Ship*, pp. 257-260
79. Muirhead, 'An Object in Stasis: Space, Place & Materiality', p. 189
80. Smith, *Coal, Steam and Ships*, pp.114 -115
81. Lambert, 'Brunel, the Navy and the Screw Propeller', pp. 30-31
82. Muirhead, 'An Object in Stasis: Space, Place & Materiality', p. 48

SOURCES AND BIBLIOGRAPHY

Primary Sources
National Archives, London
BT 41/36 /209 Australian Steam Navigation Company 20 Dec 1851
BT 107/106 London Foreign Shipping registers
BT 107/329 Liverpool Foreign Shipping registers
CO 201/458 ASNC applying for more runs from bimonthly to monthly service.
RAIL 1149/60 Half yearly and Special Reports of various companies 1836-1880
TS 45/58 Treasury Solicitors file 4 November 1852. Delay to HM Mails carried in steam ship *Melbourne*, 1852

London Metropolitan Archives
Antony Gibbs & Co Collection: Gibbs, Bright papers
CLC/B/012/MS11077 Bills of Sale of Ships 1839-1863
CLC/B/012/MS11078 Mortgages on ships

V & A Archive of Art & Design
Crace family, interior decorators: papers, 1691–1992

Brunel Institute, *SS Great Britain*, Bristol
University of Bristol Brunel collection
Brunel Letter books 5 January 1847

DM 1306/5/2-3 correspondence between Brunel and Christopher Claxton re the SS *Great Britain*, 1856

SS Great Britain Collection
Heywood Bright letters 1855
Diary of Samuel Archer
Diary of Louise Buchan
Diary of John Campbell
Diary of Olcher Fedden 1852
Diary of Allan Gilmour 1852
Diary of E. M. Grace 1863–4
Diary of Daniel Higson
Diary of Edward Towle, 1852
Diary of Rosamund D'Ouseley, Transcript from OM 74 74 State Library of Queensland, Oxley Library, Level 4 South Bank Building, South Brisbane, Queensland 4101 Australia
Diary of Robert Saddington, 1853 Transcript from MS271 Library of New South Wales, Mitchell Library, Level 4, South Bank Building, South Brisbane, Queensland 4101, Australia
Copies of Liverpool Customs Bills of Entry
Great Western Steam Ship Company AGM papers 1839–1848

Database of passengers and crew on board *Great Britain*, from Crew Agreements and passenger lists
Global Stories; family data on passengers and crew from descendents and other research

Unpublished research notes:
Malcolm Lewis, 'Ship Surgeons and the *ss Great Britain*', November 2018
Mike Hinton, 'Cholera on *Great Britain* during July 1855'
Peter Revell, 'Captain John Gray: Biographical Notes'
Peter Revell, 'Robert Bright Biographical Notes'

On-Line Collections
Ancestry.com
Passenger Lists of Vessels Arriving at New York, New York, 1820-1897. Microfilm Publication M237, 675 rolls.

NAI: 6256867. Records of the U.S. Customs Service, Record Group 36. National Archives at Washington, D.C.

HO 3: Returns of alien passengers, July 1836-December 1869, returns made of alien passengers on ships arriving at British ports as required by the Aliens Act, beginning in 1836; formerly known as Lists of Immigrants

UK and Ireland, Masters and Mates Certificates, 1850-1927 Original data: *Master's Certificates*. Greenwich, London, UK: National Maritime Museum

UK Census 1841-1881

Selected British and Irish Newspapers

Bristol Mirror
Bristol Mercury
Devizes and Wiltshire Gazette
Dublin Evening Packet
Dublin Monitor
Durham County Advertiser
Evening Post
Gloucester Journal
Hampshire Chronicle
Illustrated London News
Inverness Courier
Kentish Weekly Post
Liverpool Albion
Lloyds List
London Dispatch
London Evening Standard
Morning Post
Norwich Mercury
Staffordshire Advertiser
Sun
Taunton Courier & Western Advertiser
The Times
Worcester Journal
Yorkshire Gazette

US Newspapers
Courier and Enquirer
Herald
New York Evening Post

Canadian Newspaper
Patriot and Farmers Monthly

British Parliamentary Reports
BPP Select Committee on Causes of Shipwrecks. Report, Minutes of Evidence 1836

BPP Report on Steam Vessel Accidents to Committee of Privy Council for Trade, 1839

BPP Report from the Select Committee on Halifax and Boston Mail 1846

BPP Return of Troop Transports 1856

BPP 19th General Report of the Emigration Commissioners 1859

BPP 1852: Return of Vessels and Number of Immigrants for South Australia and Victoria 1847 to 1852.

BPP: Return of Vessels and Number of Immigrants for South Australia and Victoria 1847 to 1852

Unpublished Theses
Connor, Claire, 'Passengers, Emigrants and Modern Men: A Social History of the 1852 Voyage of SS *Great Britain* from Liverpool to Melbourne' PhD, University of Bristol, 2014

Mendonca, Sandro, 'The evolution of new combinations: drivers of British maritime engineering competitiveness during the nineteenth century', PhD, University of Sussex, 2012

Muirhead, James, 'The ss Great Britain – an Object in Stasis: Space, Place & Materiality, 1886–2015', PhD, University of Bristol 2016

Nineteenth-century Books and Diaries
Lloyd's Captains Register 1865
Lloyd's Register 1834

Post Office Directory 1837

Alexander, Colonel Sir James E. (ed), *The Albatross: A Voyage from Victoria to England* (Stirling: Charles Rogers & Co, 1863)

Anon, 'Obituary. Joshua Field (Ex-President and Vice-President), 1786-1863', *Minutes of the Proceedings of the Institution of Civil Engineers*, 23 (1864), 488-92.

Anon, 'Obituary. Thomas Richard Guppy, 1797-1882', *Minutes of the Proceedings of the Institution of Civil Engineers*, 69 (1882), 411-15.

Claxton, Christopher, *A History and Description of the Steam Ship Great Britain* (Bristol and New York, 1845).

Claxton, Christopher, *Logs of the First Voyage, Made with the Unceasing Aid of Steam, between England and America by the Great Western of Bristol* (Great Western Steamship Company, 1838).

Claxton, Christopher, *The Naval Monitor*, 2nd edn (London: 1833).

Gurner, John Augustus, *Life's Panorama: being recollections and reminiscences of things seen, things heard, things read* (Melbourne: Lothian Pub Co Pty Ltd, 1869).

Guppy, T R, 'Report' in *Minutes and Proceedings of the Institution of Civil Engineers* March 1845.

Hosken, James, *Autobiographical Sketch of the Public Career of Admiral James Hosken* (Penzance: Rodda, 1889).

Park, Thomas Murray, *Memoirs Book 2*, 1852.

Ritchie, George and Davies, Henry (eds), *The Great Britain Times* (Birkenhead: Griffith, 1866).

Tuckerman, Byard (ed), *The Diary of Philip Hone, 1828 to 1851* (New York, Dodd, Mead & Co 1889).

Bibliography

Adams, David (ed) *Letters of Rachel Henning* (Sydney, NSW: Angus & Robertson, 1963).

Armstrong, John and Williams, David M., 'The Steamboat and Popular Tourism', *Journal of Transport History*, 26 (2005), 61-77.

Armstrong, John, and Williams, David, 'The Impact of Technological Change: The Early Steamship in Britain', Vol.

47, *Research in Maritime History* (St John's, Newfoundland: International Maritime Economic Association, 2011).

Bagust, Harold, *The Greater Genius? A Biography of Sir Marc Isambard Brunel* (Hersham: Ian Allen Publishing, 2006).

Banbury, Philip, *Shipbuilders of the Thames and Medway* (Newton Abbot: David and Charles, 1971).

Brinnin, John Malcolm, *The Sway of the Gand Saloon*. 1986 edn (London: Arlington Books, 1971).

Broeze, F. J. A., 'The Cost of Distance: Shipping and the Early Australian Economy, 1788-1850', *Economic History Review*, 28 (1975), 582-97.

Broeze, Frank, 'Distance Tamed: Steam Navigation to Australia and New Zealand from the Beginnings to the Outbreak of the Great War', *Journal of Transport History*, X (1989), 1-21.

Broeze, Frank, 'Private Enterprise and the Peopling of Australasia, 1831-50', *Economic History Review*, 35 (1982), 235-53.

Brown, D. K., 'The First Steam Battleships', *The Mariner's Mirror*, 63 (1977), 327-33.

Brown, D. K., 'The First Steam Battleships', *The Mariner's Mirror*, 63 (1977), 327-33.

Brown, David K., 'Seppings, Sir Robert (1767–1840)', *Oxford Dictionary of National Biography* (Oxford University Press, 2004).

Brunel Noble, Celia, *The Brunels – Father and Son* (London: 1938).

Brunel, Isambard, *The Life of Isambard Kingdom Brunel Civil Engineer*. 2006 edn (Stroud: Nonsuch Publishing, 1870).

Bryan, Tim, *Brunel: The Great Engineer* (Hersham: Ian Allan, 2006).

Buchanan, R. A. and Doughty, M. W., 'The Choice of Steam Engine Manufacturers by the British Admiralty, 1822–1852', *The Mariner's Mirror*, 64 (1978), 327-47.

Buchanan, R. Angus, *Brunel: The Life and Times of Isambard Kingdom Brunel* (London: Hambledon and London, 2002).

Burrows, Edwin G., and Mike Wallace, *Gotham: A History of New York City to 1898* (Oxford: Oxford University Press, 1999).

Burton, Valerie C., 'Apprenticeship Regulation and Maritime Labour in the Nineteenth Century British Merchant Marine', *International Journal of Maritime History*, 1 (1989), 29-49.

Bush, G. W. A., *Bristol and Its Municipal Government, 1820-1851* (Bristol: Bristol Record Society, 1976).

Chaloner, W. H. and Henderson, W. O., 'Aaron Manby, Builder of the First Iron Steamship', *Transactions of the Newcomen Society*, 29 (1953), 77-91.

Clydesdale, Greg, 'Thresholds, Niches and Inertia: Entrepreneurial Opportunities in the Steamship Industry', *Journal of Enterprising Culture*, 20 (2012), 379-404.

Colvile, R. F., 'The Navy and the Crimean War', *Royal United Services Institution Journal*, 85 (1940), 73-78.

Corlett, Ewan, *The Iron Ship: The Story of Brunel's SS Great Britain*. 2012 edn (Bristol: ss Great Britain, 1975).

Cottrell, P. L., 'The Steamship on the Mersey, 1815-80', in *Shipping, Trade and Commerce: Essays in Memory of Ralph Davis* ed. by P. L. Cottrell and D. H. Aldcroft (Leicester: Leicester University Press, 1981), pp. 137-64.

Crompton, James (ed.), *A Journal of a Honeymoon Voyage in SS Great Britain* (SS Great Britain Project:1992)

Crossley, John, 'Death and Disease on the SS Great Britain', *Proceedings of Bristol Medico-Historical Society*, 3 (1999), 89-93.

Curtin, Emma, 'Gentility Afloat: Gentlewomen's Diaries and the Voyage to Australia, 1830–80', *Australian Historical Studies*, 26 (1995), 634-52.

de la Valette, John, 'The Fitment and Decoration of Ships', *Journal of the Royal Society of Arts, Manufacturing and Commerce*, 84 (1936), 705-26.

de Oliveira Torres, Rodrigo, 'Handling the Ship: Rights and Duties of Masters, Mates, Seamen and Owners of Ships in 19th Century Merchant Marine', *International Journal of Maritime History*, 26 (2014), 587-99.

Doe, Helen, *The First Atlantic Liner: Brunel's Great Western Steamship*. (Stroud: Amberley Publishing 2017).

Doe, Helen, 'Power, Authority and Communications: The Role of the Master and the Managing Owner in Nineteenth Century Merchant Shipping', *International Journal of Maritime History*, Vol XXV, No 1, June 2013.

Doe, Helen., 'Waiting for her Ship to come in: The Female Investor in Nineteenth Century Shipping', *Economic History Review*, Vol.63, No.1, February 2010, pp.85-106.

Doe, Helen., *Enterprising Women and Shipping in the Nineteenth Century* (Woodbridge: Boydell and Brewer, September 2009)

Dresser, Madge, 'Guppy, Sarah (*bap.* 1770, *d.* 1852)', *Oxford Dictionary of National Biography*, Oxford University Press, May 2016

Ellmers, Chris, '"This Great National Object" – the Story of the Paddlesteamer Enterprize', in *Shipbuilding and Ships on the Thames: Proceedings of Fourth Symposium*, ed. by Roger Owen (London Docklands Museum 2009), pp. 59-81.

Erickson, Arvel B., and Fr. John R. McCarthy, 'The Yelverton Case: Civil Legislation and Marriage', *Victorian Studies*, 14 (1971), 275-91.

Evans, Richard J., 'Epidemics and Revolutions: Cholera in Nineteenth-Century Europe', *Past & Present* (1988), 123-46.

Fahey, Graham, 'Peopling the Victorian Goldfields: From Boom to Bust, 1851-1901', *Australian Economic History Review*, 50 (2010), 148-61.

Farr, Grahame, *Records of Bristol Ships:1800-1838* (Bristol: Bristol Record Society, 1950).

Farr, Grahame, *Shipbuilding in the Port of Bristol, Maritime Monographs and Reports* (Greenwich, London: National Maritime Museum 1977).

Fox, Stephen, *The Ocean Railway: Isambard Kingdom Brunel, Samuel Cunard and the Revoluntionary World of the Great Atlantic Steamships* (London: Harper Collins, 2004).

Freeman, Mark, Robin Pearson, and James Taylor, 'Law, Politics and the Governance of English and Scottish Joint-Stock Companies, 1600–1850', *Business History*, 55 (2013), 633-49.

Gillin, Edward John, '"Diligent in Business, Serving the Lord": John Burns, Evangelicalism and Cunard's Culture of Speed, 1878–1901', *Journal for Maritime Research*, 14 (2012), 15-30.

Goold-Adams, Richard, *The Return of the Great Britain* (London: Weidenfield and Nelson, 1976).

Goold-Adams, Richard, 'The SS *Great Britain* and Its Salvaging', *Journal of the Royal Society of Arts,* 119 (1971), 234-48.

Gower, L. C. B., 'The English Private Company', *Law and Contemporary Problems,* 18 (1953), 535-45.

Griffiths, Denis, Andrew Lambert, and Fred Walker, *Brunel's Ships* (London: Chatham Publishing, 1999).

Griffiths, Denis, *Brunel's Great Western* (Wellingborough: Patrick Stephens, 1985).

Haines, Robin, and Ralph Shlomowitz, 'Immigration from the United Kingdom to Colonial Australia: A Statistical Analysis', *Journal of Australian Studies,* 16 (1992), 43-52.

Haines, Robin, 'Medical Superintendence and Child Health on Government-Assisted Voyages to South Australia in the Nineteenth Century', *Health and History,* 3 (2001), 1-29.

Hamilton, C. I., 'Three Cultures at the Admiralty, c.1800–1945: Naval Staff, the Secretariat and the Arrival of Scientists', *Journal for Maritime Research,* 16 (2014), 89-102.

Harcourt, Freda, 'British Oeanic Mail Contracts in the Age of Steam, 1838-1914', *Journal of Transport History,* IX (1988), 1-18.

Harcourt, Freda, 'The High Road to India: The P&O Company and the Suez Canal, 1840-1874', *International Journal of Maritime History,* XXII (2010), 19-72.

Hart, Douglas, 'Sociability and "Separate Spheres" on the North Atlantic: The Interior Architecture of British Atlantic Liners, 1840-1930', *Journal of Social History,* 44 (2010), 189-212.

Hays, J. N., 'Lardner, Dionysius (1793–1859)', *Oxford Dictionary of National Biography* (Oxford University Press, 2004); online edn, Oct 2007 [http://o-www.oxforddnb.com.lib.exeter.ac.uk/view/article/16068,

Henning, G. R., 'Competition in the Australian Coastal Shipping Industry During the 1880s', *International Journal of Maritime History,* 5 (1993), 157-73.

Hope, Ronald, *A New History of British Shipping* (London: John Murray, 1990).

Humphreys, David, 'Competition in the Merchant Steamship Market, 1889-1914', *The Mariner's Mirror,* 99 (2013), 429-43.

Hyde, Francis E., *Cunard and the North Atlantic, 1840-1973* (London: MacMillan Press, 1975).

Kaukiainen, Yrjo 'Shrinking the World: Improvements in the Speed of Information Transmission, c. 1820-1870', *European Review of Economic History*, 5 (2001), 1-28.

Kelly, Andrew and Kelly, Melanie, *Brunel, in Love with the Impossible : A Celebration of the Life, Work, and Legacy of Isambard Kingdom Brunel* (Bristol: Bristol Cultural Development Partnership, 2006).

Kennedy, Greg, 'Maritime Strength and the British Economy, 1840-1850', *The Northern Mariner/Le Marin du nord,* VII (1997), 51-69.

Kennerley, Alston, 'Early State Support of Vocational Education: The Department of Science and Art Navigation Schools, 1853-63', *Journal of Vocational Education and Training,* 52 (2000), 211-24.

Kennerley, Alston, 'Nationally Recognised Qualifications for British Merchant Navy Officers, 1865-1966', *International Journal of Maritime History,* XIII (2001), 115-35.

Kennerley, Alston, 'Stoking the Boilers: Firemen and Trimmers in British Merchant Ships, 1850-1950', *International Journal of Maritime History,* XX (2008), 191-220.

Klovland, Jan Tore, 'New Evidence on the Fluctuations in Ocean Freight Rates in the 1850s', *Explorations in Economic History,* 46 (2009), 266-84.

Laakso, Seija-Riitta, *Across the Oceans: Development of Overseas Business Information Transmission 1815-1875* (Helsinki: Studia Fennica, 2007).

Lambert, Andrew D., 'Preparing for the Long Peace: The Reconstruction of the Royal Navy 1815–1830', *The Mariner's Mirror,* 82 (1996), 41-54.

Lambert, Andrew, 'Preparing for the Russian War: British Strategic Planning, March 1853-March 1854', *War & Society,* 7 (1989), 15-39.

Lambert, Andrew, 'Brunel, the Navy and the Screw Propeller' in Griffiths, Denis, Andrew Lambert, and Fred Walker, *Brunel's Ships* (London: Chatham Publishing, 1999).

Lambert, Andrew, 'Captain Sir William Symonds and the Ship of the Line: 1832–1847', *The Mariner's Mirror,* 73 (1987), 167-79.

Lambert, Andrew, 'John Scott Russell: Ships, Science and Scandal in the Age of Transition', *The International Journal for the History of Engineering & Technology,* 81 (2011), 60-78.

Lambert, Andrew, 'Woolwich Dockyard and the Early Steam Navy, 1815 to 1852', in *Shipbuilding and Ships on the Thames: Proceedings of Fourth Symposium,* ed. Roger Owen (London Docklands Museum 2009), pp. 82-96.

Langley, John G., *Steam Lion: A Biography of Samuel Cunard* (Halifax, Nimbus, 2006)

Large, David (ed), *The Port of Bristol 1848-1884* (Bristol: Bristol Record Society, 1984)

Laughton, J. K., 'Hosken, James (1798–1885)', rev. Andrew Lambert, *Oxford Dictionary of National Biography,* Oxford University Press, 2004

Leggett, Don, and Davey, James, 'Introduction: Expertise and Authority in the Royal Navy, 1800–1945', *Journal for Maritime Research,* 16 (2014), 1-13.

Lewis, Michael, *The Navy in Transition, 1814-1864* (London: Hodder & Stoughton, 1965).

Lin, Chih-lung, 'The British Dynamic Mail Contract on the North Atlantic: 1860–1900', *Business History,* 54 (2012), 783-97.

MacDonagh, Oliver, 'The Regulation of the Emigrant Traffic from the United Kingdom 1842-55', *Irish Historical Studies,* 9 (1954), 162-89.

Mackay, David, 'Desertion of Merchant Seamen in South Australia, 1836–1852: A Case Study', *International Journal of Maritime History,* 7 (1995), 53-73.

Macleod, Christine, Jeremy Stein, Jennifer Tann, and James Andrew, 'Making Waves: The Royal Navy's Management of Invention and Innovation in Steam Shipping, 1815–1832', *History and Technology,* 16 (2000), 307-33.

Maenpaa, Sari, 'Galley News: Catering Personnel on British Passenger Liners, 1860-1938', *International Journal of Maritime History,* XII (2000), 243-60.

Maenpaa, Sari, 'Women Below Deck: Gender and Employment on British Passenger Liners, 1860-1938', *Journal of Transport History*, 25 (2004), 57-74.

Markovits, Stefanie, 'Rushing into Print: "Participatory Journalism" During the Crimean War', *Victorian Studies*, 50 (2008), 559-86.

Marsden, Ben, and Crosbie Smith, *Engineering Empires: A Cultural History of Technology in Nineteenth-Century Britain* (London: Palgrave, 2005).

Mathew, Kate, 'The Female Middle Class Emigration Society: Governesses in Australia: A Failed Vision?', *Journal of Australian Colonial History*, 14 (2012), 107-30.

McDermid, Jane, 'Home and Away: A Schoolmistress in Lowland Scotland and Colonial Australia in the Second Half of the Nineteenth Century', *History of Education Quarterly*, 51 (2011), 28-48.

McDonald, John, and Ralph Shlomowitz, 'Contract Prices for the Bulk Shipping of Passengers in Sailing Vessels, 1816–1904: An Overview', *International Journal of Maritime History*, 5 (1993), 65-93.

McMurray, H. Campbell, 'Ship's Engineers: Their Status and Position on Board, c. 1830-65', in *West Country Maritime and Social History* ed. by Stephen Fisher (Exeter: University of Exeter, 1980).

Merrill, James M., 'British-French Amphibious Operations in the Sea of Azov, 1855', *Military Affairs*, 20 (1956), 16-27.

Milburn, R. G., 'The Emergence of the Engineer in the British Merchant Shipping Industry, 1812–1863', *International Journal of Maritime History*, 28 (2016), 559-75.

Milne, Graeme J., *Trade and Traders in Mid-Victorian Liverpool: Mercantile Business and the Making of a World Port* (Liverpool: Liverpool University Press, 2000).

Morgan, Kenneth, 'The Bristol Chamber of Commerce and the Port of Bristol, 1823-1848', *International Journal of Maritime History*, 18 (2006), 55-77.

Myers, Janet C., 'Performing the Voyage Out: Victorian Female Emigration and the Class Dynamics of Displacement', *Victorian Literature and Culture*, 29 (2001), 129-46.

Neill, Elizabeth, *Fragile Fortunes; The Origins of a Great British Merchant Family* (Wellington: Halsgrove, 2008)

Noble, Celia Brunel, *The Brunels – Father And Son* (London: 1938).

O'Byrne, William R., *A Naval Biographical Dictionary* (London: John Murray, 1849).

Owen, J. R., 'The Post Office Packet Servcie, 1821-37: Development of a Steam-Powered Fleet', *The Mariner's Mirror,* 88 (2002), 155-75.

Palmer, Sarah, 'Experience, Experiment and Economics: Factors in the Construction of Early Merchant Steamers', *Proceedings of the Atlantic Canada Shipping Project* (1977), 233-49.

Palmer, Sarah, 'The Most Indefatigable Activity: The General Steam Navigation Company, 1824-50', *Journal of Transport History,* 3 (1982), 1-22.

Payton, P., *Making Moonta – The Invention of Australia's Little Cornwall* (University of Exeter, Exeter, 2007).

Payton, P., *The Cornish Miner in Australia; Cousin Jack Down Under* (Dyllansow Truran, Redruth, 1984).

Pearson, Robin, 'Shareholder Democracies? English Stock Companies and the Politics of Corporate Governance During the Industrial Revolution', *The English Historical Review,* 117 (2002), 840-66.

Penn, Chris, 'The Medical Staffing of the Royal Navy in the Russian War, 1854-6', *The Mariner's Mirror,* 89 (2003), 51-58.

Pietsch, Tamson, 'A British Sea: Making Sense of Global Space in the Late Nineteenth Century', *Journal of Global History,* 5 (2010), 423-46.

Pietsch, Tamson, 'Bodies at Sea: Travelling to Australia in the Age of Sail', *Journal of Global History,* 11 (2016), 209-28.

Quinlan, Michael, 'Balancing Trade with Labour Control: Imperial/ Colonial Tensions in Relation to the Regulation of Seamen in the Australian Colonies, 1788–1865', *International Journal of Maritime History,* 9 (1997), 19-56.

Quinn, Paul, 'Wrought Iron's Suitability for Shipbuilding', *The Mariner's Mirror,* 89 (2003), 437-61.

Quinn, Paul, 'I K Brunel's Ships – First among Equals?', *The International Journal for the History of Engineering & Technology*, 80 (2010), 80-99.

Quinn, Paul, 'Macgregor Laird, Junius Smith and the Atlantic Ocean', *The Mariner's Mirror*, 92 (2006), 486-97.

Rankin, Stuart, 'William Evans, Shipbuilder of Rotherhithe and His Steamships', *Shipbuilding on the Thames and Thames-built ships; Proceedings of a second symposium* (2004), 47-76.

Robinson, Howard, *Carrying British Mail Overseas* (London: George Allen & Unwin, 1964).

Rolt, L. T. C., *Isambard Kingdom Brunel*, 1st edition 1957 (London: Penguin, 1976).

Scally, Robert Scally, 'Liverpool Ships and Irish Emigrants in the Age of Sail', *Journal of Social History*, 17 (1983), 5-30.

Scholl, Lars U., 'The Loss of the Steamship *President*: A Painting by German Artist Andreas Achenbach', *The Northern Mariner/ Le Marin du nord*, 15 (2005), 53-71.

Sloan, Edward W., 'The Baltic Goes to Washington: Lobbying for a Congressional Steamship Subsidy, 1852', *The Northern Mariner/Le Marin du nord*, 5 (1995), 19-32.

Sloan, Edward W., 'The First and Very Secret International Steamship Cartel', in *Global Markets: The Internationalization of Sea Transport Industries Since 1850* ed. by David R Starkey and Gelina Harlaftis (St John's, Newfoundland: IMEHA 1998), pp. 29-52.

Smallman-Raynor, Matthew and Andrew D. Cliff, 'The Geographical Spread of Cholera in the Crimean War: Epidemic Transmission in the Camp Systems of the British Army of the East, 1854–55', *Journal of Historical Geography*, 30 (2004), 32-69.

Smith, Crosbie and Anne Scott, '"Trust in Providence": Building Confidence into the Cunard Line of Steamers', *Technology and Culture*, 48 (2007), 471-96.

Smith, Crosbie, '"Imitations of God's Own Works": Making Trustworthy the Ocean Steamship', *History of Science*, 41 (2003), 379-426.

Smith, Crosbie, *Coal, Steam and Ships: Engineering, Enterprise and Empire on the Nineteenth-century Seas* (Cambridge: Cambridge University Press, 2018).

Smith, Crosbie, Ian Higginson and Phillip Wolstenholme, '"Avoiding Equally Extravagance and Parsimony": The Moral Economy of the Ocean Steamship', *Technology and Culture*, 44 (2003), 443-69.

Smith, Crosbie, '"This Great National Undertaking"; John Scott Russell, the Master Shipwrights and the Royal Mail Steam Packet Company', in *Re-Inventing the Ship: Science, Technology and the Maritime World, 1800-1918*, ed. by Don Leggett and Richard Dunn (Basingstoke: Ashgate, 2012), pp. 25-52.

Stammers, M. K., 'Letters from the *Great Britain* 1852', *The Mariner's Mirror*, 62 (1976), 284-86.

Stammers, Mike, *The Emigrant Clippers to Australia: The Black Ball Line, Its Operation, People and Ships, 1852-1871* (Barnoldswick: Milestone Research).

Staniforth, Mark, 'Deficiency Disorder: Evidence of the Occurrence of Scurvy on Convict and Emigrant Ships to Australia 1837 to 1839', *The Great Circle*, 13 (1991), 119-32.

Starkey, David R., 'The Industrial Background to the Development of the Steamship', in *The Advent of Steam: The Merchant Ship before 1900*, ed. R Gardiner (London: Conway Maritime Press, 1993).

Steel, Frances, 'Women, Men and the Southern Octopus: Shipboard Gender Relations in the Age of Steam, 1870s–1910s', *International Journal of Maritime History*, 20 (2008), 285-306.

Stevens, John, *Bristol Politics in the Age of Peel, 1832-1847* (Bristol: Avon Local History & Archaeology, 2014).

Strachan, Hew, 'Soldiers, Strategy and Sebastopol', *The Historical Journal*, 21 (1978), 303-25.

Suchard, Philippe (1827), *Mein Besuch Amerikas im Sommer 1824 (My visit to America in the summer of 1824)* (Switzerland: Aarau; Voegtli, M. 2003)

Trotter, Lesley, *The Married Widows of Cornwall: The Story of the Wives Left behind by Emigration* (Humble History Press, 2018)

Tyrrell, Margot & Alex, 'The Hosken Family Papers: A Naval Genealogy', *The Mariner's Mirror*, 74 (1988), 273-82.

Williams, D., 'James Silk Buckingham: Sailor, Explorer and Maritime Reformer', in Stephen Fisher, *Studies in British*

Privateering, Trading and Seamen's Welfare, 1775-1900 (Exeter, University of Exeter, 1987), pp 99-120

Williams, David M., and John Armstrong, '"One of the Noblest Inventions of the Age": British Steamboat Numbers, Diffusion, Services and Public Reception, 1812–c. 1823', *Journal of Transport History*, 35 (2014), 18-34.

Winter, Alison, '"Compasses All Awry": The Iron Ship and the Ambiguities of Cultural Authority in Victorian Britain', *Victorian Studies*, 38 (1994), 69-98.

Young, Chris, *The Incredible Journey: The SS Great Britain Story 1970-2010* (Bristol: SS Great Britain Trust, 2010).

LIST OF ILLUSTRATIONS

A large number of the illustrations in this book are held at the SS Great Britain Trust in Bristol. They hold the nationally recognised collection of items relating to Isambard Kingdom Brunel. Within this there are distinct collections, among them the Clive Richards Collection, accepted under the Cultural Gifts Scheme by HM Government from Clive Richards OBE DL and allocated to the SS Great Britain Trust in 2017.

32 The *Great Britain* being escorted on her transatlantic return passage (SS Great Britain Trust)

33 Richard Goold-Adams meeting Jack Hayward at Lulsgate Airport (SS Great Britain Trust)

34 Towing the *Great Britain* round the Avon's Horseshoe Bend in 1970 (SS Great Britain Trust)

35 Ewan Corlett, Jack Hayward and Lord Strathcona with the ship's hull in the background (SS Great Britain Trust)

36 Safely back in her original dock (SS Great Britain Trust)

37 From the bows

38 Floating in her glass sea (SS Great Britain Trust)

39 *Great Britain* lit up at night (SS Great Britain Trust)

INDEX

Guppy, Thomas 16, 24-26,
30, 34, 35, 42-46, 50, 84
Hall, Samuel 18
harbourmaster, Port
Stanley 181, 183
Harman, H S, engineer 26,
46
Haskard, Sir Cosmo 187, 194
Hawes, William 102
Hayward, Sir Jack 190, 193,
196-7, 199
Healey, Denis, politician 187
Henniker-Heaton, Sir
Herbert 181-183
Henning, Rachael 154
Hilhouse, George 19
Hill, Charles and son 188,
193
Hill, Richard 188
Hone, Philip 59
Hosken, Captain Jame, *see*
masters
Humphreys, Francis 18, 19,
26, 27, 168
Humphreys, Edward 168
Humphreys, engine 18,19, 26

Indian mutiny 116-117
Inman Line 93
Inness, George, artist 92
insurance 23, 75, 76, 81, 88
iron shipbuilding benefits 17
Jones, George 30
Jones, Samuel, Chief Justice,
New York 59
Kane, Elisha Kent,
explorer 54
Kellock and Co 172
King, William Rufus
Devane 70-71

Kingstown, Ireland 50
Kortum, Karl 183, 185, 186,
195

Lamport, Charles,
shipbuilder 168
Lardner, Dr Dionysius 58
launch, floating out 36-37
lifeboats 42, 74, 79
Liverpool Underwriters 65,
68, 81, 89, 119
Lloyd's Register 23, 76
Longworth, Teresa *see*
Yelverton
Lunel & Co, shipbuilders 19
Lynch, Margaret 131

Mackay, Thomas 108, 123,
169
mail contracts 15, 29, 31, 56,
75, 80, 102-103, 105, 107,
125
masters
Chapman, Charles 167,
170
Gray, John 99, 100, 114,
118, 126, 127, 136-142
Hosken, James 38, 46, 47,
50, 58, 83-84
Kerr, Francis 174, 175
Mathews, Barnard 38,
91-99, 136-7
Morris, James 174
Robertson, Peter 139, 170
Stap, Henry 174, 175-77
matrons 145-151
Maudslay, Sons & Field 13,
17, 30, 34
Maze, Peter 19, 27
MacIver, Liverpool agents 71

Also available from Amberley Publishing

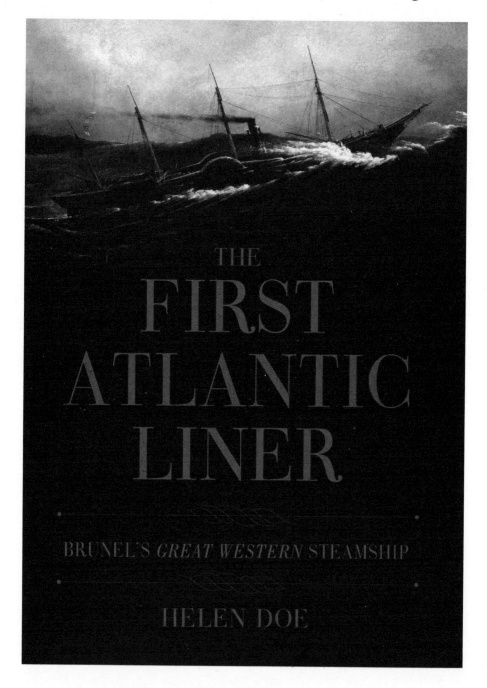

THE
FIRST
ATLANTIC
LINER

BRUNEL'S *GREAT WESTERN* STEAMSHIP

HELEN DOE

Available from all good bookshops or to order direct
Please call **01453–847–800**
www.amberley-books.com